T0186990

Eduqas
Physics
A Level

Revision Workbook 2

Gareth Kelly

Iestyn Morris

Nigel Wood

Published in 2022 by Illuminate Publishing Limited, an imprint of Hodder Education, an Hachette UK Company, Carmelite House, 50 Victoria Embankment, London EC4Y ODZ

Orders: Please visit www.illuminatepublishing.com or email sales@illuminatepublishing.com

British Library Cataloguing in Publication Data

A catalogue record for this book is available from the British Library

ISBN 978-1-912820-65-8

Printed by CPI Group (UK)

02.22

The publisher's policy is to use papers that are natural, renewable and recyclable products made from wood grown in sustainable forests. The logging and manufacturing processes are expected to conform to the environmental regulations of the country of origin.

Editor: Geoff Tuttle

Design and layout: Nigel Harriss

Cover image: © Shutterstock / Aqua Images

Acknowledgements

The authors would like to thank Adrian Moss for guiding us through the early stages of production and the Illuminate team of Eve Thould, Geoff Tuttle and Nigel Harriss for their patience and meticulous attention to detail. We are also indebted to Dr Sarah Ryan for her eagle eyes in spotting mistakes and inconsistencies and for her many insightful suggestions.

Contents

How to use this book

What this book contains

The A Level Physics course consists of three components each assessed with a separate exam. This book covers Component 3, **Light, Nuclei and Options**, which is assessed by a 2 hour 15 minute exam.

Component 3 has fourteen areas of study, of which ten are core areas and four are options. You should study only one option.

Each area of study has its own section in this book and there you will find:

- a **concept map**, which displays how the different concepts within the area of study are related to one another and to other areas of study;
- a set of **graded questions**, with space for your answers, which are designed to test the content of the area of study in a way that is similar to the examination;
- a **question and answers** section containing one or two examples of exam-style questions, with answers from two students, Rhodri and Ffion (who produce answers of different standards), together with examiner marks and discussion.

The next section consists of a Component 3 **practice paper**. The final section consists of **model answers** to all the graded questions and the practice paper.

How to use this book effectively

This book can be used exclusively for revision, in which case you will work your way gradually through the questions as you revise each area of study. Alternatively, it can be used regularly as you are learning A Level Physics, with each set of questions providing an end-of-area check/test. Your teacher might even like to use it as a homework book because it contains 14 sets of end-of-area questions as well as the practice papers.

Probably the worst method of revising physics is just reading your notes or your textbooks, a technique that will simply send you to sleep. Only a small fraction of the information will be retained. Regardless of the quality of writing of the notes/textbooks, there is an inherent, soporific effect to reading notes. One way of combating this tendency of losing concentration is to make your own notes as you read through the notes/textbook. Even so, you will be preparing yourself only for exam questions in which you reproduce learned material. These, so-called Assessment Objective 1 (AO1) questions form only 30% of the exams. The higher-level skills required for AO2 and AO3 require different revision techniques. See pages 5 to 7 for further information on assessment objectives.

You will find that the best way of revising is to answer exam-type questions. As in many fields, it is practice that makes perfect in physics, and you should practise as many past papers and questions within this book as you can.

You will, inevitably, come across questions that seem difficult. Should you not be able to answer these questions, it is time to visit your notes/textbook, but this time you will be doing so with a purpose to help you concentrate. Should the question still leave you stumped, have a look at the answers in the back of the book. If the answer seems unintelligible, it is time to ask your teacher or your fellow students for an explanation of how and why the answer is correct. Your teacher may also point out that different answers are creditworthy, especially in discussion questions.

So, answering the questions and analysing the model answers will do more for your exam preparation than just reading notes or making revision timetables (however colourful)!

Assessment objectives

You need to demonstrate expertise in answering examination questions in different ways. Because it's physics, you will need to use maths in most questions. Some questions require expertise in practical skills. Besides these two categories are the assessment objectives. There are three of them.

Assessment Objective 1 (AO1)

AO1 questions are ones in which you need to:

demonstrate knowledge and understanding of scientific ideas, processes, techniques and procedures

This objective accounts for 30% of the marks in the three component papers.

The official description in bold (**demonstrate ... procedures**) sounds far more complicated and highfalutin than is necessary. Essentially, these are the marks that can be obtained without too much thinking. This category covers:

- Recall of definitions, laws and explanations from the specification
- Inserting appropriate data into equations
- Deriving equations where required in the specification
- Describing experiments from the **specified practicals**.

In general, no judgement or new thinking is required. One can still learn a definition or a law without fully understanding it, so it is a good idea to use the Eduqas terms and definitions (T&D) booklet to help you memorise these. Defined T&D from the booklet are printed in **bold** in the concept map for each topic in this book.

Examples of AO1 questions

1. Explain what is meant by the wavelength of a longitudinal wave. [2]

Good answer (full marks 2/2): It is the minimum distance between points (on different wavefronts) that are oscillating in phase.

Bad answer (0/2): It is the minimum distance between two peaks or two troughs.

The *bad answer* is incorrect because longitudinal waves have compressions and rarefactions, not peaks and troughs.

2. Explain the historical significance of Young's double slit experiment. [2]

Good answer (2/2): Since Newton's time, light was considered to be made of particles. Young's experiment proved that light had wave properties (diffraction and interference).

Bad answer (0/2): Young's experiment proved that light was made up of photons, which disproved earlier theories.

The *bad answer* does not include bad physics but has jumped forward in time by almost exactly 100 years. Photons and wave particle duality were discovered around 1900, whereas Young's experiment was 1802. Note that this is not an issues question for AO3 marks because *the historical importance of Young's experiment* is stated on the syllabus.

3. Show that the relationship between the half-life and the decay constant of a radioactive nucleus is $\lambda T_{1/2} = \ln 2$, where λ is the decay constant and $T_{1/2}$ is the half-life. [2]

Good answer (3/3): $N = N_0 E^{-\lambda t}$ but one half-life is when the number of nuclei drop from N_0 to $\frac{N_0}{2}$.

Hence, $\frac{N_0}{2} = N_0 E^{-\lambda T_{1/2}} \rightarrow \frac{1}{2} = e^{-\lambda T_{1/2}}$.

Now, we take logs of both sides: $\ln \frac{1}{2} = -\lambda T_{1/2} \rightarrow \ln 2 = \lambda T_{1/2}$ QED

Bad answer (0/3): Starting from the equation in the data booklet, $\lambda = \frac{\ln 2}{T_{1/2}}$. I just have to rearrange this for $\ln 2 = \lambda T_{1/2}$. Is this really worth 3 marks???

The bad answer comes from a candidate who has not noticed that 3.6(h) of the syllabus states *the derivation and use of* $\lambda = \frac{\ln 2}{T_{1/2}}$. This is why this derivation should be learned and corresponds to AO1 marks.

Assessment Objective 2 (AO2)

AO2 questions are ones in which you need to:

apply knowledge and understanding of scientific ideas, processes, techniques and procedures:
1. **in a theoretical context**
2. **in a practical context**
3. **when handling qualitative data**
4. **when handling quantitative data.**

Here, the key words are 'apply knowledge'. The application of knowledge is required in theoretical, practical, qualitative and quantitative contexts. Theoretical just means some idealised context made up by the examiner. Practical means that the data have apparently come from a real experiment (although the data are usually made up by the examiner). *Qualitative* means without numbers and calculations whereas *quantitative* means the opposite (i.e. with numbers and calculations).

Note that application of knowledge here can also include analysis of data even though 'analyse' also appears in AO3 (see page 7). Generally, if you are told what type of analysis to carry out, these will be AO2 skills. If the question is more open-ended and you must choose the analysis methods yourself, the question will be classified as AO3. Questions relating to AO2 are the most common type and account for 45% of the marks on the papers. Note that all calculations must be mainly AO2 marks: we have seen that inserting data into an equation is classed as AO1, but any manipulation of an equation, such as changing the subject, and the production of a final answer is AO2 (unless it is AO3).

Examples of AO2 questions

1. The speed of light in a certain medium is 1.423×10^8 m s^{-1}. Calculate the refractive index of the medium.
    [2]

Good answer (2/2): $n = \dfrac{c}{v} = \dfrac{3.00 \times 10^8 \text{ m s}^{-1}}{1.423 \times 10^8 \text{ m s}^{-1}} = 2.1$

Bad answer (1/2): $n = \dfrac{c}{v} = \dfrac{3.00 \times 10^8}{1.423 \times 10^8} = 2.10$

The bad answer is penalised 1 mark for incorrect rounding (the actual answer is 2.1082...).

2. Calculate the activity of 1.26 μg of carbon-14, whose half-life is 5730 years.
    [4]

Good answer (4/4): decay constant, $\lambda = \dfrac{\ln 2}{T_{1/2}} = \dfrac{\ln 2}{5730 \times 365 \times 24 \times 60 \times 60} = 3.84 \times 10^{-12}$ s^{-1}

Number of mol, $n = \dfrac{1.26 \times 10^{-6}}{14}$; number of nuclei, $N = nN_A = \dfrac{1.26 \times 10^{-6}}{14} \times 6.02 \times 10^{23}$

∴ Activity, $A = \lambda N = 3.84 \times 10^{-12} \times \dfrac{1.26 \times 10^{-6}}{14} \times 6.02 \times 10^{23} = 208$ kBq

Bad answer (1/4): decay constant $m = \dfrac{\ln 2}{T_{1/2}} = \dfrac{\ln 2}{5730} = 1.21 \times 10^{-4}$

Activity $A = \lambda N = 1.21 \times 10^{-4} \times 1.26 \times 10^{-6} = 1.5 \times 10^{-10}$ Bq

The bad answer has not converted the time unit but still gets the first mark for using the equation. Mass is substituted into the activity equation rather than number of nuclei, so there are definitely no more marks awarded! (This is quite a common mistake made by candidates who can't obtain the number of nuclei.)

3. A long solenoid has length 85.0 cm and 6800 turns. Calculate the B-field at its centre when the current is 5.00 A.
    [2]

Good answer (2/2): $B = \mu_0 nI = 4\pi \times 10^{-7}$ H m$^{-1} \times \dfrac{6800}{0.850 \text{ m}} \times 5.00 \text{ A} = 0.0503$ T

Bad answer (0/2): $B = \mu_0 nI = 4\pi \times 10^{-7} \times 6800 \times 5.00 = 0.043$

In the bad answer, the student does not realise that is the number of turns **per unit length**, so the first mark, for correctly using an equation, is not gained. The second mark is unavailable because the first mark was lost for incorrect physics.

Assessment Objective 3 (AO3)

AO3 questions are ones in which you need to:

analyse, interpret and evaluate scientific information, ideas and evidence, including in relation to issues, to:
1. **make judgements and reach conclusions;**
2. **develop and refine practical design and procedures.**

These questions account for 25% of the marks in the three component papers.

The verbs *analyse*, *interpret* and *evaluate* are all appropriate as these are, indeed, what you will have to do. Most of these AO3 marks will concentrate on the first point – *judgements* and *conclusions*. The context will often be similar to one of the specified practicals with realistic data. Your analysis may well include analysing graphs to make numerical conclusions. You might have to evaluate the quality of the data and your conclusions. In some questions, you are given a statement and have to determine whether or not (or to what extent) it is true. There are usually several ways of getting a sensible answer: you must choose one and structure your answer carefully. Other questions relate to the second part of the AO3 statement – develop and refine practical design and procedures. Usually, these questions are based on imperfections in the data and how you could improve the procedure or the apparatus to obtain better data. To answer these questions, you will need to read them carefully because there will be a clue (perhaps right at the start) as to what went wrong.

Another type of question is based on the part of the statement '*including in relation to issues*'. The '*issues*' include: risks and benefits; ethical issues; how new knowledge is validated; how science informs decision making. Try and make sensible comments; the mark scheme will allow for many approaches and the marks will be quite attainable – approach the questions like a politician: have a view. Every theory paper has one issues question.

Examples of AO3 questions

1. Discuss the quality of data obtained by Gaz for the relationship between magnetic force and current in a wire. [4]

Good answer (4/4): A line of best fit can be drawn through all the error bars, which indicates that there are no anomalous points. The best fit line is a straight line, in agreement with theory ($F = BIl$ and $F \propto I$). However, the steepest and least steep lines do not straddle the origin which they would do if force and current were proportional. I believe Gaz might have forgotten to press the 'zero' button on the digital balance just before starting the experiment, giving a small intercept on the force axis.

Bad answer (1/4): The best fit line is straight and passes close to all the points. The graph isn't perfect because the best fit line doesn't quite pass through the origin.

The bad answer gets the mark for the straight line best fit but, although the answer suggests a good understanding, it requires more detail.

2. The gradient of the graph of log(count rate) against log(distance) for gamma radiation was measured as −1.8 ± 0.3. Discuss whether this value is consistent with the expected value for this experiment. [2]

Good answer (2/2): The expected gradient from theory is −2 (due to the inverse square law). The value obtained means that the actual gradient could be anything from −1.5 to −2.1. This range includes the expected value of −2 and so is consistent with theory.

Bad answer (1/2): We expect a gradient of −2, so the measured value is pretty close.

The bad answer gets the mark for knowing the correct expected gradient but hasn't discussed the uncertainty in the value.

Preparing for the examinations

Examination mark schemes

When examiners write questions for A level exams, they also provide mark schemes containing details of how they are to be marked. For an example of a question and its mark scheme, see page 87. You'll notice that each part of the question is covered, with details of the sort of answer required for each mark. The mark scheme also contains information about the assessment objectives and any marks which count towards the mathematical and practical skills on the paper – in this question there are no practical skills. The question is quite heavy on AO3, however.

Let's look at this mark scheme in detail.

Part (a) is a one-mark question which requires both and identification and an explanation. Notice the expression 'or equiv' (*equivalent*) which means that the examiner will look for other ways of explaining which consist of correct physics. All the markers are current or retired physics teachers, so they know how to interpret this.

Part (b) is a typical AO3 question in which you have to draw a conclusion based on data. Notice that the marks are given for the reasons and are only given if the basic conclusion is correct, that it is an example of a strong interaction.

Part (c) is a piece of bookwork, which you are expected to know. Hence it is AO1.

Parts (e)(i)–(iii) are AO2 and include application with calculations. They are really just one question, but they are divided this way to help you pick your way through the different ideas.

In part (e)(iv) notice the letters *ecf*. These stand for *error carried forward*. You have previously calculated the total energy and need to halve it to get the mark here. If your previous calculation gave the wrong answer then you can still pick up marks for using this value. This rule is generally applied even when the mark scheme doesn't say so explicitly.

The marking

Now have a look at some of Rhodri's and Ffion's marked answers to this question. Notice that the examiner has put in ticks and crosses, where the mark has been given or withheld. You'll see also some annotations by the examiner. If you get a mark by ecf, the examiner will write this – see Rhodri's answer to (e)(iv).

In Ffion's (d) answer, the examiner has written 'not enough.' This shows that it needed a more specific reason.

Another common annotation is *bod* – see Rhodri's answer to part (c). This stands for *benefit of the doubt*. Rhodri's statement about needing high energy is not explicitly tied in with repulsion but the examiner thought there was enough of a 'hint.'

Notice also that Rhodri used a valid method of calculation which was different from that in the mark scheme in (e)(i). The experienced physics examiner noticed that it was sound physics and awarded the marks.

The Component 3 examination paper

The time allowed is 2 hours 15 minutes. The paper is divided into two sections, **A** and **B**. **Section A** consists of 100 marks set on the core content of Component 3. As there are 10 areas of study you might expect to see 10 questions with 10 marks on each. While the examiners will allocate a fair spread of marks to each area of study, there are four things (other than randomness of distribution) that arise to mess up this beautifully symmetric system:

1. Practical content: You can expect 20% of this section examination (20 marks) to be based on experimental analysis. This usually means that one (or two) of the questions will be based on the specified practicals for this component. This could be a description of the method, error analysis, graphs and conclusions. These are often the longest questions on the paper. Examiners try to ensure that the different component papers cover different practical skills, so once you have taken Components 1 and 2, you should have an idea of what to expect in Component 3.

2. **Quality of extended response** (QER): This is a 6-mark question with a lot of lines for writing and maybe some space for diagrams, too. These tend to be AO1 marks and so rely on you learning the basic physics required to answer the question. This, however, is only part of the problem. Not only must you put the required information down on paper, but you must also do so in a logical, well-presented format, employing good language skills. The penalty for poor English is generally only 1 or 2 marks but the penalty for not knowing the relevant physics is 6 marks! Describing a specified practical is a common QER question.

3. **Synoptic content**: Any of the areas of study of Components 1 and 2 can be combined with a Component 3 area of study to make a more difficult question, e.g. conservation of momentum might be combined with the energy released in a decay. Hence you need to make sure that you are still up to speed on the Component 1 and 2 content when you take this paper.

4. **Issues:** See the earlier AO3 section.

Section B (the options)

There are four 20-mark questions: one for each of the optional topics. There are no specified practicals in the options and the QER question will be in Section A. The examiner will try to ensure a fair spread of marks across the different aspects of each option but with only 20 marks to play with, each year's paper will contain different selections from the option content. 50% of the marks in each question will be mathematical and 25% will be AO3.

Key command words and phrases in examination questions

These are the words or phrases which let you know what sort of answer is expected – there are quite a few to look out for:

State: Just provide a statement without an explanation.

Example: State the meaning of A and Z in the $_{Z}^{A}X$ symbol for a nuclide.

Answer: Z is the atomic number (or proton number, or number of protons in the nucleus); A is the atomic number (or nucleon number or total number of protons and neutrons in the nucleus).

Define: You need to provide a statement which is close to (or equivalent to) that which appears in the Eduqas Terms and Definitions booklet.

Example: Define binding energy of a nucleus.

Answer: It's the energy that has to be supplied in order to dissociate the nucleus into its constituent nucleons.

Explain what is meant by (or explain the meaning of ...): This is slightly more complicated than define and can mean a couple of things:

1. Sometimes it just means the same as 'define'.
   Example: Explain what is meant by the binding energy of a nucleus.

   Answer: [Exactly the same as above.]

2. Sometimes it's a definition with a number included.
   Example: Explain what is meant by the statement, 'The activity of a β⁻ radioactive source is 1.6 MBq'.

   Answer: The source gives out 1.6×10^{6} β⁻ particles (or electrons) per second.

Explain the difference (between two things): This is two definitions in disguise because if you define both things you have automatically explained the difference between them.

Example: Explain the difference between the magnetic flux and flux linkage through a coil.

Answer: The flux of a magnetic field of strength (flux density), B, at an angle θ to a coil of area A is $\Phi = AB\cos\theta$. The flux linkage is the flux multiplied by the number of turns of the coil.

Describe: Provide a brief description but no explanation is required.

Example: Describe how the binding energy per nucleon depends upon the atomic number, A, of nuclei.

Answer: As the nucleon number increases (from 1) the binding energy per nucleon increases (with a spike at ⁴He) up to a maximum at around ⁵⁶Fe and then gradually decreases.

Explain ... (some statement): Sometimes this requires a reason-by-reason logical argument. Example: Explain briefly why the fission of ²³⁵U releases energy.

Answer: The binding energy per nucleon of ²³⁵U is greater than that of the fission products.

Suggest ... (or suggest a reason ...): Although not a common command word, this can produce some questions that are difficult to answer. These will often be AO3 marks, appearing at the end of a question requiring evaluation skills.

Example: The measured count rate at 10 cm from the radioactive source is much less than that expected from its calculated activity. Suggest a reason.

Possible answers: The radioactive source might be emitting α particles which have a range of less than 10 cm (in air) / the emissions from the middle of the source are absorbed before they are able to emerge.

Calculate or **determine**: The aim is to obtain the correct answer (along with the correct unit, if required by the mark scheme). With this command word, the correct answer will obtain full marks without the workings. However, you are advised strongly to show your working as marks are available for this even if the answer is wrong.

Example: Calculate the magnetic field strength 1.2 cm from a long straight wire due to a 6.0 A current in the wire.

Answer: $B = \dfrac{\mu_0 I}{2\pi r} = \dfrac{4\pi \times 10^{-7} \text{ H m}^{-1} \times 6.0 \text{ A}}{2\pi \times 1.2 \times 10^{-2} \text{ m}} = 1.0 \times 10^{-4}$ T

[Note that you do not have to put units in the calculation – but you do in your answer!]

Compare: Not a common command word but you ought to do what it says on the tin – compare the things it says to compare in the question.

Example: The decay constant of nuclide 1 is 10 × that of nuclide 2. Compare (a) the half-lives and (b) the activities of samples of the two nuclides with the same number of atoms.

Answer: (a) Half-life of 2 = 10 × half-life of 1. (b) Activity of 1 = 10 × activity of 2.

Evaluate: You will be required to make a judgement, e.g. whether a statement is correct or wrong, or to decide whether data are good or a final value is accurate.

Justify: This is sometimes used in a very similar manner to the word 'determine' when AO3 marks are being examined, e.g. justify whether or not Amrita was correct in stating that the 2.00 V reading was anomalous.

Discuss: This can often be a command word in the 'issues' question. In general, you will not go far wrong if you make a couple of points in favour of the discussion issue, a couple against it and then draw some sort of a sensible conclusion.

Common exam mistakes

1. **Not converting the given numbers correctly:** Planetary distances are usually in km while wire radii are in mm. Resistors can be in Ω, kΩ, MΩ and these have to be converted to the correct powers of ten. There are other common conversions such as changing diameter to radius when using area or volume formulae. All these can give rise to simple mistakes which do not show a poor understanding of physics. Such mistakes are not penalised more than one mark most of the time. Nonetheless, these are probably the most common mistakes committed by physics students.

2. **Not reading the question carefully enough**: This usually results in not answering the question that was asked – either by answering a different question altogether or by missing part of the question. The most common parts of questions that are omitted are those that do not have dotted lines for you to answer on, e.g. adding to diagrams. Pay particular attention to these short parts of questions. Other common missed questions are ones that have an **and** condition in the question itself, e.g. calculate the magnitude **and** the direction. One or other part of the question may have been forgotten in the answer.

3. **Not understanding equations properly**. This often involves substituting wrong values into equations – an unforgivable sin! In kinematics equations, for instance, u and v are often mixed up. You shouldn't really have to use the data booklet; you should know the equations intimately and only check it from time to time to ensure that you recollect them correctly. How do you ensure that you don't misunderstand an equation? Practise, practise, practise!

4. **Not knowing the basic terms and definitions** (a surprisingly common cause of loss of marks). There is an Eduqas booklet full of these – you should know everything within its covers.

5. **Forgetting to square a value in the equation**. This happens most often with the kinetic energy equation – the equation $E = \frac{1}{2}mv^2$ is written correctly but then the candidate forgets to square the velocity on the calculator. Or the converse: forgetting to square root the answer when using the same equation to calculate the velocity!

6. **Not planning the structure before answering the QER** (and extended explanations). Too many QER responses are rambling and unstructured. This is easily remedied by spending a moment to plan and structure your answer. Using short sentences tends to help, too.

7. **Not matching the correct corresponding values in a calculation**. By far the most common mistake here is with electrical circuits: current, pd and resistance, e.g. a pd and a current will be combined to obtain a resistance ($R = V / I$) but the current and pd do not match – the pd is for one resistor and the current for another.

Component 3: Light, nuclei and options

Section 1: The nature of waves

Topic summary

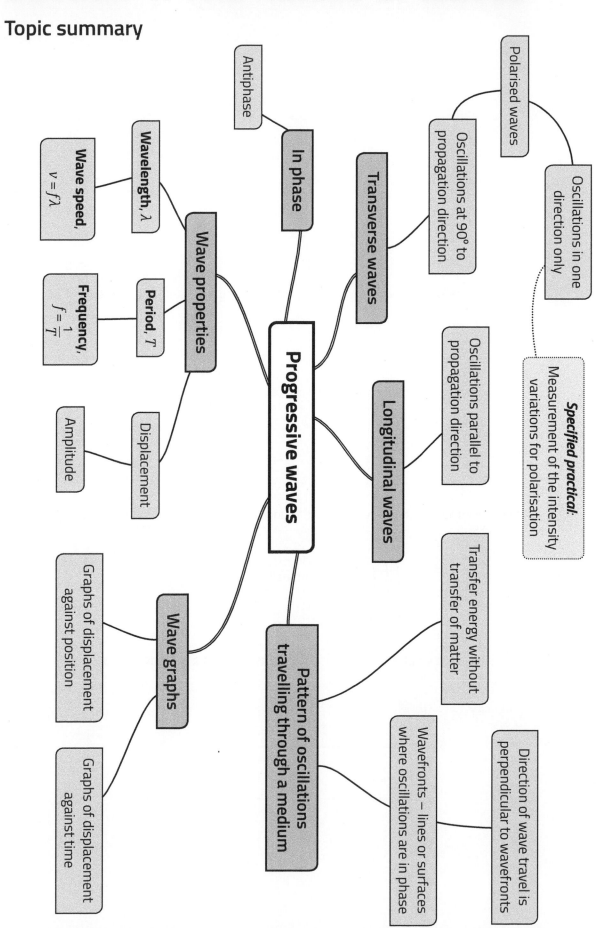

Q1 Energy can be transferred by moving fuel from one place to another. Explain briefly how energy transfer using progressive waves is different. [2]

....

....

....

Q2 Explain the difference between transverse and longitudinal waves and give an example of each. [4]

....

....

....

....

....

....

....

Q3 (a) Explain what is meant when light is described as being *polarised in one direction*. [1]

....

....

(b) Explain what is meant when light is described as being *partially polarised*. [2]

....

....

....

Q4 Unpolarised light of intensity 1.00 W m^{-2} is incident normally upon a slowly rotating polaroid and then a detector.

(a) Explain why the graph shown of intensity against rotation angle is to be expected. [2]

....

....

....

(b) Cheryl rotates a polaroid in a beam of light and obtains the following graph of intensity against polaroid angle.

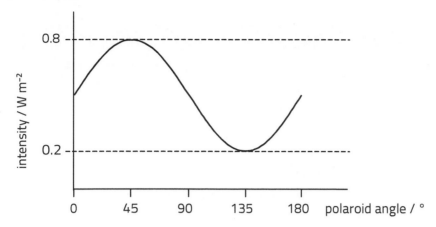

She suggests that the incident light beam has a total intensity of 1.0 W m⁻², with 60% polarised in one direction and the remainder unpolarised. Evaluate her suggestion. [4]

....

....

....

....

....

....

....

Q5 Explain how an unpolarised light source and two polaroids can be used to investigate the polarisation of light and describe the expected observations. [6 QER]

....

....

....

....

....

....

....

....

....

....

....

....

....

Q6 A sinusoidal wave travels in one direction on a long string.

A

3.04 m

At time $t = 0$, point A has a maximum displacement.

(a) Indicate a point on the string oscillating:
(i) in phase with A (label it B);
(ii) in antiphase with A (label it C). [2]

(b) A graph of the displacement of point A with respect to time is shown.

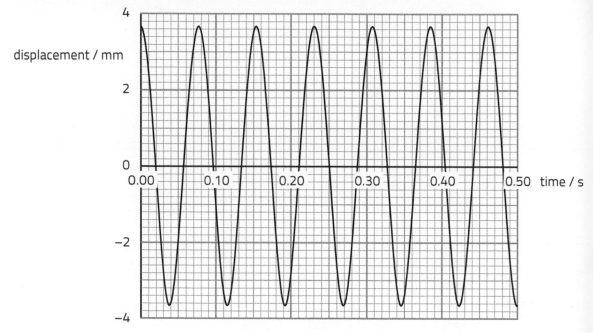

Calculate

(i) the wavelength, [1]

....

....

(ii) the amplitude, [1]

....

....

(iii) the speed of the wave. [3]

....

....

....

....

....

....

Q7   The diagram is a snapshot of circular wavefronts that are peaks on the surface of water.

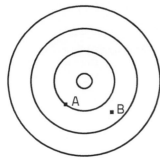

(a) Indicate on the diagram:

   (i)   the direction of propagation for point A;   [1]

   (ii) a distance corresponding to a wavelength (label it λ);   [1]

   (iii) a point oscillating in phase with point B (label it C);   [1]

   (iv) a point oscillating in antiphase with point B (label it D).   [1]

(b) Joanna notices that 25 wavefronts pass point B in 8.0 s and that the distance between the first and fourth wavefronts is 18.0 cm. Calculate the speed of the wave.   [3]

Q8   An earthquake strikes Wales with its epicentre at the Liberty Stadium in Swansea. The earthquake produces both longitudinal (P) and transverse waves (S). The P waves from the earthquake travel at a speed of 6.2 km s^{-1} and the S waves travel at a speed of 3.7 km s^{-1}. Both the S and P waves have a frequency of 8.9 Hz.

(a) Calculate the wavelengths of the P and S waves.   [3]

(b) In Bala, a North Wales town, a seismogram reveals that there is a delay of 16.3 s between the arrival of the longitudinal waves and the transverse waves. Calculate the distance between Swansea and Bala.   [3]

Q9 Waves travel along the surface of water in a ripple tank. The wavefronts and direction of propagation of the waves can be seen in the diagram.

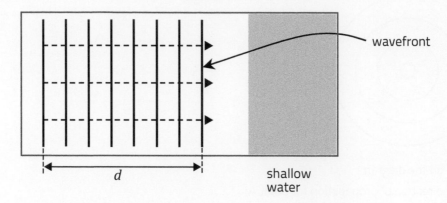

(a) State the relationship between the direction of propagation, the line of wavefronts and the direction of oscillation of the water surface. [2]

....

....

....

(b) Gerallt measures the distance, d, shown in the diagram as 14.6 cm and also counts 20 waves passing a certain point in 4.7 s. Calculate the speed of the waves. [4]

....

....

....

....

....

....

....

....

....

(c) The waves now pass to the shallow water where the propagation speed is less. Gerallt states that the frequency of the waves stays the same and their wavelengths shorten. Evaluate Gerallt's conclusions. [3]

....

....

....

....

....

Question and mock answer analysis

Q&A 1

(a) Explain the difference between the terms *displacement* and *amplitude* for a wave. [2]

(b) State a definition of the term *wavelength* that applies to both transverse and longitudinal waves. [2]

(c) The displacement of waves is obtained at varying distances from the source at time t = 0 using a photograph. A graph of displacement against distance is then produced.

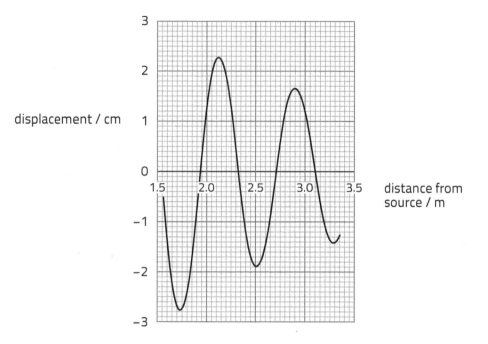

(i) Describe briefly the variation in amplitude and explain why this occurs. [2]

(ii) The frequency of oscillation of the wave is 44 Hz. Sketch a graph of the variation of the displacement of the wave with time at 2.9 m from the source on the grid below.

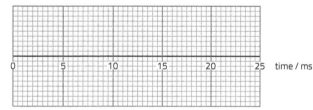

[4]

(iii) Calculate the speed of the waves. [3]

(iv) The intensity of the wave is proportional to the amplitude squared. Evaluate whether or not the variation in intensity follows the inverse square law $\left(\text{intensity} \propto \dfrac{1}{\text{distance}^2} \right)$ [4]

What is being asked

Parts (a) and (b) are almost standard definitions but have a slight twist. Nonetheless, they are still AO1 marks. Part (c)(i) requires a simple description of the results and an explanation for this description. Although evaluation of results is usually AO3, this is more of a quick piece of analysis and a standard explanation for the amplitude drop. Parts (c)(ii) & (iii) are also based on analysis and calculation and hence are AO2 marks. Part (c)(iv), on the other hand, involves quite tricky AO3 marks because checking for the inverse square law requires a good understanding and strategy.

Mark scheme

Question part			Description	AOs			Total	Skills	
				1	2	3		M	P
(a)			Displacement is the distance from the equilibrium position [1] Amplitude is the maximum displacement [1]	2			2		
(b)			The (minimum) distance between two adjacent points [1] Oscillating in phase (but not on the same wavefront) [1]	2			2		
(c)	(i)		Amplitude decreases with distance [1] Due to wave spreading out in 2D/3D OR accept due to resistive forces/damping [1]		2		2		
	(ii)		Period calculated or implied (22.7 ms) [1] Amplitude is 1.65 cm. Accept 1.6 – 1.7 (cm) [1] y-axis labelled (numbers, title, unit) [1] Reasonable freehand sinusoid; correct phase (cos wave), period and amplitude in plot [1]		4		4	3	
	(iii)		Obtaining wavelength = 0.75 ± 0.05 (m) [1] Using equation $v = f\lambda$ [1] Answer = 33 (m s^{-1}) [1]	1	1 1		3	2	
	(iv)		Selecting 1 set of appropriate data, e.g. dist = 1.75, A = 2.75, dist = 3.25, A = 1.45 etc. [1] Realising amplitude$^2 \propto \dfrac{1}{\text{distance}^2}$ [1] Choosing an appropriate method e.g. checking Ad = constant OR A^2d^2 = constant OR calculating a second amplitude/distance using $A = \text{constant}/d$ [1] Result and valid conclusion e.g. 4.76, 4.71 quite close (allow ecf) [1]			4	4	4	
Total				5	8	4	17	9	

Rhodri's answers

(a) Displacement is the instantaneous height of the wave (needs more, no bod) whereas amplitude is the biggest value that this can take. ✓

MARKER NOTE
Height of the wave is not good enough for the displacement especially if the wave is longitudinal. However, 'the biggest value it can take' is equivalent to the maximum in the mark scheme.

1 mark

(b) The wavelength is the distance between consecutive crests or compressions. ✓ bod

MARKER NOTE
Rhodri has been crafty and has come up with an answer that almost satisfies the demand of the question. He does not know this standard definition but his answer satisfies the condition for the first mark, i.e. 'the distance between consecutive crests or compressions' is equivalent to 'The distance between two adjacent points'.

1 mark

(c) (i) The amplitude is clearly decreasing over time and so this is damped oscillations. ✓ bod

MARKER NOTE
Rhodri has completely misunderstood what is going on. He thinks that the amplitude is decreasing over time when it is decreasing with distance from the source. Nonetheless, his point about damping hits the 2nd marking point and is awarded the 2nd mark with bod.

1 mark

(ii) $T = \frac{1}{f} = \frac{1}{44}$ ✓

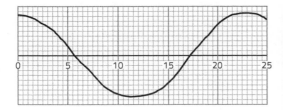

✓ bod for amplitude (no scale)

✓

MARKER NOTE
Here, Rhodri's final graph is better than Ffion's but he obtains fewer marks because he does not explain clearly what he is doing. The 1st mark is deserved for the period, especially when the graph is examined (leaving the answer as a fraction is not a good idea). The 2nd mark is awarded with a very generous bod because the amplitude is implied from the graph (even though no y-axis scales are shown). The 3rd mark cannot be awarded because the y-axis is unlabelled. The 4th mark is deserved for a good plot but still requires bod because the actual amplitude is implied rather than shown clearly to be correct.

3 marks

(iii) 36 m s^{-1} No bod

MARKER NOTE
Rhodri's risky tactic of not showing his workings has backfired here. He has gained zero marks even though his answer is possibly correct (see Ffion's answer). His speed of 36 m s^{-1} suggests a wavelength of 0.82 m which is outside tolerance and the examiner is left with no option but to award zero marks. Had Rhodri written 35 m s^{-1} he might well have received full marks but, as it appears, there is no evidence of any correct workings.

0 marks

(iv) Are you `avin' a laugh? What does inverse square law mean? My teacher hasn't taught me this! I reckon the statement is correct because it usually is. If you look at the numbers, when the amplitude goes down to 74% (2.3 to 1.7) the distance increases ✓ from 2.1 to 2.9 (no bod) which is sort of the same ratio (2.1 is 72% of 2.9). ✓ (bod)

MARKER NOTE
First, the examiner has to smile at Rhodri's silly comments and then ignore them (as long as they are not serious enough to be reported). Then, the examiner realises that Rhodri is very close to obtaining full marks even though it is possible that he doesn't know what he is doing. The 1st mark is obviously deserved but the 2nd mark is definitely not awardable. The 3rd mark is difficult to judge because Rhodri is, essentially, checking that the amplitude and distance are inversely proportional but the examiner has decided against awarding bod. The 4th mark is awarded because the correct conclusion is reached (just) and this method is valid even though Rhodri has not explained (or understood) this method. Note that this is an excellent example of perseverance leading to marks – many candidates would have left this part blank and received nothing.

2 marks

Total	8 marks /17

Ffion's answers

(a) Displacement is a vector and is defined as the distance from the equilibrium position ✓ and amplitude is the maximum displacement. ✓

MARKER NOTE
Ffion's answer is even better than the mark scheme because she has added that the displacement is a vector.

2 marks

(b) The wavelength is the distance between the two nearest points ✓ bod oscillating in phase. ✓

MARKER NOTE
Ffion's answer has received the benefit of doubt for the 1st mark but any examiner who understands English would have awarded the mark. Ffion has replaced the word 'adjacent' with 'nearest' which should be acceptable to any examiner. Note that the word 'minimum' is in brackets in the mark scheme. This means that the word 'minimum' is an optional extra and is not compulsory.

2 marks

(c) (i) The amplitude is decreasing as you get further and further away from the source ✓ – this is the inverse square law. No bod

MARKER NOTE
Ffion's answer is arguably better but only scores the same mark as Rhodri. She has given a simple description of the amplitude variation for the 1st mark but the second part of her answer is contained in question c(iv). Hence the examiner must decide whether she knows what she is talking about or has she just copied the details from part c(iv). As a general rule, the examiners can't award marks for details that appear in the questions and so Ffion is stuck on 1 mark

1 mark

(ii) To calculate the period $T = \frac{1}{f} = 0.023$ s ✓

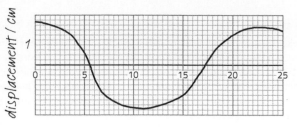

The amplitude is 1.65 cm ✓
  ✓ bod amplitude looks slightly out at the top
  ✓ bod only one number on axis

MARKER NOTE
Ffion's answer is excellent and contains better explanations than Rhodri's but she still needs bod to obtain full marks. Her y-axis has a title and correct unit but only has one number. It is true that one number is enough to define all the other numbers on the axis but it was a risky thing to do – a stricter examiner might have withheld the mark. Ffion's other slight mistake is that although the minimum value of displacement is quite accurate (around −1.7 cm) her maximum value is slightly out (around 1.55 cm). However, this is not an easy graph to sketch and the examiner has rightly allowed bod for this mark.

4 marks

(iii) wavelength = 0.8 m ✓
    Speed = 44 × 0.8 = 35.2 m/s ✓✓

MARKER NOTE
Ffion's wavelength is slightly large but within tolerance. This correctly leads to a speed of 35.2 m s⁻¹ and she deservedly gains full marks.

3 marks

(iv) The amplitude at distance 1.75 is 2.85 cm while the amplitude at 3.275 is 1.45 cm. ✓
    If $I \propto 1/d^2$ then $I = k / d^2$
    Using first data $k = I\,d^2 = 1.75 \times 2.85^2 = 14.2$
    Using 2nd data $k = I\,d^2 = 3.275 \times 1.45^2 = 6.9$
    Hence, I conclude that the intensity does not follow the inverse square law. ✓ ecf

MARKER NOTE
Ffion has selected better data than Rhodri (separated by more distance) and has read from the graph to more precision. Unfortunately, there is only 1 mark for this and she is not rewarded for going the extra mile on this occasion. Ffion then forgets an important detail in the question – that the intensity is proportional to the amplitude squared. She must then lose the 2nd and 3rd marks. However, she can gain the last mark by ecf as is stated in the mark scheme.

2 marks

| Total | 14 marks /17 |

Section 2: Wave properties

Topic summary

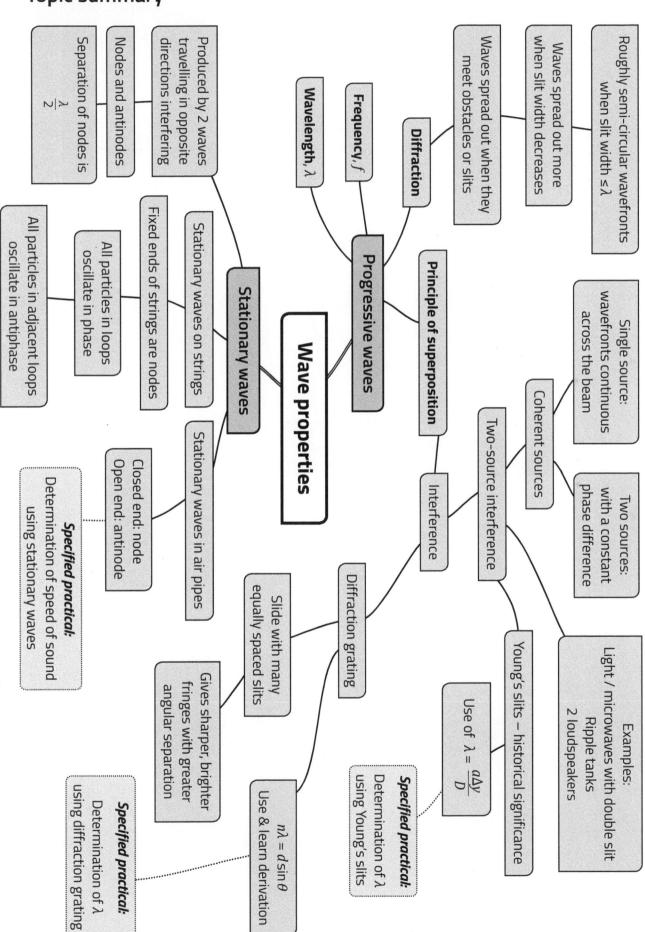

Wave properties

Progressive waves

Stationary waves

Wavelength, λ

Frequency, f

Diffraction
- Waves spread out when they meet obstacles or slits
- Waves spread out more when slit width decreases
- Roughly semi-circular wavefronts when slit width $\leq \lambda$

Produced by 2 waves travelling in opposite directions interfering

Nodes and antinodes
- Separation of nodes is $\frac{\lambda}{2}$

Stationary waves on strings
- Fixed ends of strings are nodes
- All particles in loops oscillate in phase
- All particles in adjacent loops oscillate in antiphase

Stationary waves in air pipes
- Closed end: node
  Open end: antinode

Specified practical: Determination of speed of sound using stationary waves

Principle of superposition

Interference

Two-source interference
- Coherent sources
  - Single source: wavefronts continuous across the beam
  - Two sources: with a constant phase difference
- Examples: Light / microwaves with double slit
  Ripple tanks
  2 loudspeakers
- Young's slits – historical significance
  - Use of $\lambda = \dfrac{a\Delta y}{D}$

Specified practical: Determination of λ using Young's slits

Diffraction grating
- Slide with many equally spaced slits
- Gives sharper, brighter fringes with greater angular separation
- $n\lambda = d\sin\theta$
  Use & learn derivation

Specified practical: Determination of λ using diffraction grating

Q1 *Diffraction* is a phenomenon that occurs in waves. Explain briefly what is meant by *diffraction*. [2]

....

....

Q2 Describe what happens to the diffraction pattern when light of wavelength 600 nm is shone upon a single slit whose width is increased gradually from 300 nm to 6 μm. [3]

....

....

....

....

Q3 (a) The image shows a two-source interference pattern for water waves in a ripple tank. Explain how this pattern occurs. [3]

....

....

....

....

....

....

(b) The path difference between waves from the two sources is $\frac{1}{2}\lambda$ at point A. State what the path difference is:

(i) for point B .... [1]

(ii) for point C .... [1]

(c) The two sources of waves in the ripple tank are *coherent*. State what is meant by the term *coherent*. [1]

....

....

Q4 (a) State the *principle of superposition.* [2]

....

....

....

(b) (i) Explain what is meant by the terms *constructive* and *destructive interference.* [3]

....

....

....

....

(ii) Two waves of equal amplitude but different frequencies meet. Their time-varying displacements are shown in the graph. Use the principle of superposition to obtain the net displacement at times 0 s, 0.35 s, 0.5 s, 0.65 s and 1.0 s. Hence sketch the resulting waveform on the graph. [4]

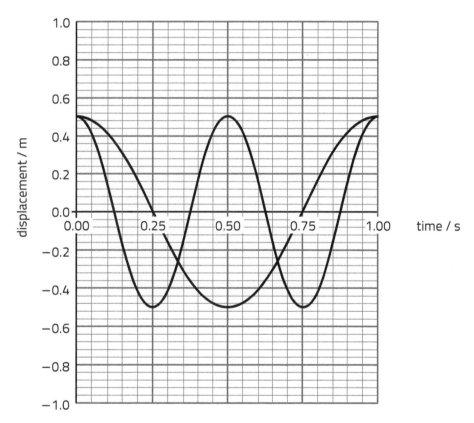

Q5 (a) State the historical significance of Young's double slit experiment. [1]

....

....

(b) Young's double slit experiment is an experiment that can be carried out in a school laboratory using a laser and double slit. Explain how the experiment should be carried out and how a suitable graph will lead to a value of the wavelength of the laser light. [6 QER]

....

....

....

....

....

....

....

....

....

....

....

....

....

....

....

....

....

....

....

Q6 Two-source interference is observed using microwaves. The microwave source is the same distance from the slits S_1 and S_2. So is the point P.

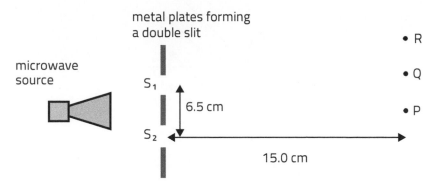

(a) Explain why a maximum signal is observed when a microwave detector is placed at point P. [3]

....

....

....

....

....

(b) Q is the first point above P where a minimum signal is detected and R is the first point above P where a maximum signal is detected.

State and explain the values of $S_2Q - S_1Q$ and $S_2R - S_1R$ in terms of the wavelength, λ, of the microwaves. [2]

....

....

....

(c) (i) The wavelength of the microwaves is 2.8 cm.

Use the equation:
$$\lambda = \frac{a\Delta y}{D}$$

to calculate the distance between point P and point Q. [2]

....

....

....

(ii) Evaluate whether or not the equation $\lambda = \frac{a\Delta y}{D}$ is a good approximation for the calculation in part (i). [2]

....

....

....

....

Q7 (a) Derive the equation for a diffraction grating. You may add to the diagram. [3]

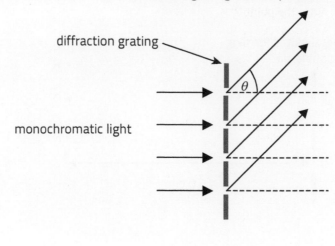

diffraction grating

monochromatic light

(b) In an experiment to measure the wavelength of laser light using a diffraction grating, Meinir obtains the following results (the $n = -1, 0, +1$ dots are shown):

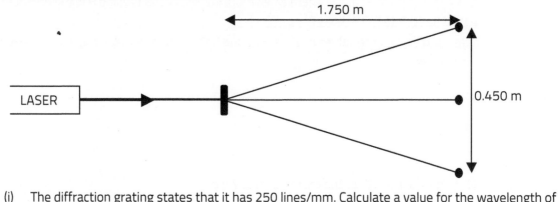

LASER

1.750 m

0.450 m

(i) The diffraction grating states that it has 250 lines/mm. Calculate a value for the wavelength of the laser light. [3]

(ii) Calculate the total number of bright dots produced by this diffraction grating when used with this laser. [3]

Q8 Meurig carries out an experiment to measure the speed of sound in air using the following apparatus.

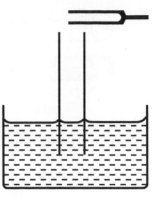

(a) Show the first harmonic in the diagram. [1]

(b) Show that the speed of sound, c, is related to the length, l, of the 1st harmonic by the relationship:

$$c = 4lf$$

where f is the frequency of the tuning fork. [2]

....

....

....

(c) Complete Meurig's table of results and discuss the accuracy of his results given that the true value of the speed of sound is 342 ms^{-1} at the temperature of the laboratory. [4]

Frequency / Hz	Length / cm	Speed / m s^{-1}
256	31.2	319
288	27.5	
320	24.6	315
384	20.1	
427	17.9	306
480	15.7	301

....

....

....

....

(d) Rachel suggests that Meurig should have drawn a straight-line graph and used the gradient to obtain the speed of sound. Evaluate to what extent Rachel is correct. [4]

....

....

....

....

....

....

....

Question and mock answer analysis

Q&A 1

(a) Explain the difference between progressive and stationary waves in terms of energy, amplitude and phase. [4]

(b) (i) The metal walls inside a microwave oven reflect microwaves. Suggest why stationary waves are produced inside a microwave oven. [2]

(ii) Brynley melts a bar of chocolate in a microwave oven with no rotating table and notices that the chocolate bar melts in positions that are separated by 6.1 cm. Calculate the frequency of the microwaves. [3]

(c) Amrita carries out an experiment to investigate the polarisation of a microwave source and adopts the following set-up.

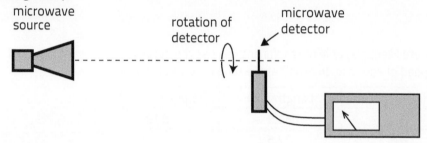

As the detector is rotated, the signal starts from a maximum and drops to zero when the detector has been rotated 90°.

Explain these observations. [4]

What is being asked

There are many, many questions that an examiner can ask within this rather large and popular topic (for examiners). This question is about stationary waves, progressive waves and microwaves. The last part of the question is slightly synoptic because it relates back to Section 3.1 – The nature of waves, and polarisation in particular. Part (a) is a standard question asking about the differences between progressive and stationary waves. Although slightly tricky, this is straight off the specification and the marks are then classified as AO1. Part (b)(i) is asking about how stationary waves are formed in a slightly unusual context and hence an analysis of what is occurring is required – these then are AO2 marks. Part (b)(ii) is a two-part calculation based on stationary waves. These are mainly AO2 marks but the simple use of the wave formula is only considered to be AO1. The final part (c) requires a slightly tricky explanation and is based on polarisation of microwaves. Here, candidates are required to make judgements and reach conclusions in a novel situation, which elevates these marks to AO3 skills.

Mark scheme

Question part			Description	AOs			Total	Skills	
				1	2	3		M	P
(a)			Progressive waves transfer energy, stationary waves do not [1]						
			Amplitude is constant (or decreases from source) for progressive but varies from max (antinode) to zero/min (node) in stationary [1]						
			Phase is constant within loop but antiphase for adjacent loops for stationary [1]	4			4		
			Phase gradually increases/changes with distance for progressive [1]						
(b)	(i)		Reflected waves interfere [1]		2		2		
			Antinodes – constructive OR nodes – destructive [1]						

Component 3 Practice questions

(ii)		Distance between antinodes [accept nodes] = $\lambda/2$ or implied [1]		1				
		Use of $c = f\lambda$ e.g. $f = \dfrac{3 \times 10^8}{0.122}$ [1]	1			3	2	
		Final answer = 2.46 GHz [1]		1				
(c)		Microwave source is polarised [1] Source & detector initially aligned giving max [1] At 90° waves from source cannot be detected [1] Explanation of either: • why microwave source is polarised OR • why detector detects only one direction of polarisation [1] e.g. electric field/current goes back and forth in one direction only (in both source and detector)			4	4		
Total			5	4	4	13	2	

Rhodri's answers

(a) Progressive waves transfer energy ✓ [bod], have a constant amplitude (but stationary varies) and the phase is continually changing.

MARKER NOTE

Rhodri's response to the energy aspect makes no mention of stationary waves not transferring energy – this is very risky. However, the examiner has awarded the mark on this occasion because the second part of the explanation is implied. Rhodri cannot gain the amplitude mark because he makes no mention of nodes and antinodes. The 3rd and 4th marks cannot be awarded either because 'the phase is continually changing' doesn't clearly mention the waves to which it refers and wouldn't be enough even if it mentioned progressive waves because there is no mention of distance from the source.

1 mark

(b) (i) As the microwaves bounce back and forth in the `popty ping` they interfere ✓ with each other and cause nodes and antinodes.

MARKER NOTE

Rhodri's answer is quite good and he has provided a bit of entertainment for the examiner by using the Welsh slang for a microwave oven (popty-ping). He obtains the 1st mark but cannot obtain the 2nd mark because he has not mentioned constructive/destructive interference.

1 mark

(ii) Using $c = f\lambda$,

$$f = \frac{c}{\lambda} = \frac{3.0 \times 10^8}{6.1} \checkmark$$

$$= 49.2 \text{ MHz } \times$$

MARKER NOTE

Rhodri cannot obtain the 1st mark because he believes the separation of nodes/antinodes to be a whole wavelength but he can obtain the 2nd mark for the use of the equation. There is no ecf applicable to obtain the last mark here. Notice that he has kept the distance in cm but this would have been penalised in the last mark, which he has already lost!

1 mark

(c) At the start the polaroids ✗ are aligned and you get a maximum signal. As you rotate the second polaroid ✗, the signal drops gradually to zero at 90° when you have cross-polaroids. ✓[ecf]

MARKER NOTE

Rhodri's answer is a reasonable attempt but he is talking about two polaroids when there are none in this question. He loses the 1st mark because he does not mention the source. He also loses the 2nd mark because he has mentioned neither the detector nor the source. The 3rd mark is awarded through ecf (as would have been agreed in the markers' conference) but the last mark is extremely difficult and cannot be awarded.

1 mark

Total   **4 marks /13**

Ffion's answers

(a) Stationary waves do not transfer energy whereas progressive do ✓ (without transferring matter). Wave amplitude drops off as inverse square for progressive waves but varies from max to zero between antinodes and nodes ✓. For a progressive wave, points lag more and more the further they are from the source ✓ but the phase is always constant inside a loop for a stationary wave.

MARKER NOTE
Ffion's energy explanation is fine and obtains the mark. Her statement regarding the amplitude of both types of waves also merits the mark. However, note that the amplitude of waves does not drop off as inverse square — it is the intensity that tends to drop off as inverse square (this is not penalised here because it is extra detail that is not required). Ffion's explanation of the phase of a progressive wave is a rare example of a correct answer but she stops short of providing enough information regarding the phase of a stationary wave — she omitted to mention that adjacent loops are in antiphase.

3 marks

(b) (i) Reflected waves will interfere ✓ with waves travelling in the opposite direction leading to areas of constructive interference (nodes) and destructive interference (antinodes). In the oven cavity there could be a 3D grid of nodes and antinodes because of the reflections off all 6 walls.

MARKER NOTE
Ffion's answer is exemplary but she has mixed up the nodes and antinodes and cannot obtain the 2nd mark. Her statement regarding the 3D pattern, although awesome, cannot be rewarded because it is not on the mark scheme.

1 mark

(ii) Wavelength = 6.1 × 2 = 12.2 cm ✓

$$f = \frac{c}{\lambda} = \frac{3.0 \times 10^8}{12.2} \checkmark$$
$$= 2.46 \times 10^7 \text{ Hz}$$

MARKER NOTE
Ffion clearly earns the first two marks but cannot obtain the final mark because she has forgotten to convert the wavelength from cm to m. This is a common slip but unusual for a very good candidate like Ffion.

2 marks

(c) First, the source must be emitting polarised waves for this effect to be observed ✓. Presumably, the detector only detects one direction of polarisation and no polaroid is needed. At the start, the source and detector are aligned and a max signal ✓ is obtained. This drops gradually to zero when the detector is rotated 90° because the detector then is at 90° to the polarisation of the source ✓, so the source has no component in the direction that the detector can detect.

MARKER NOTE
Ffion has made a superb attempt at this difficult question and clearly merits the first three marks. Her comment *'Presumably, the detector only detects one direction of polarisation and no polaroid is needed'* is insightful but falls short of the explanation required by this tough mark scheme. Her explanation that, at 90°, one polarisation has no component in the other direction is also superb but cannot be rewarded.

3 marks

| Total | 9 marks /13 |

Section 3: Refraction of light

Topic summary

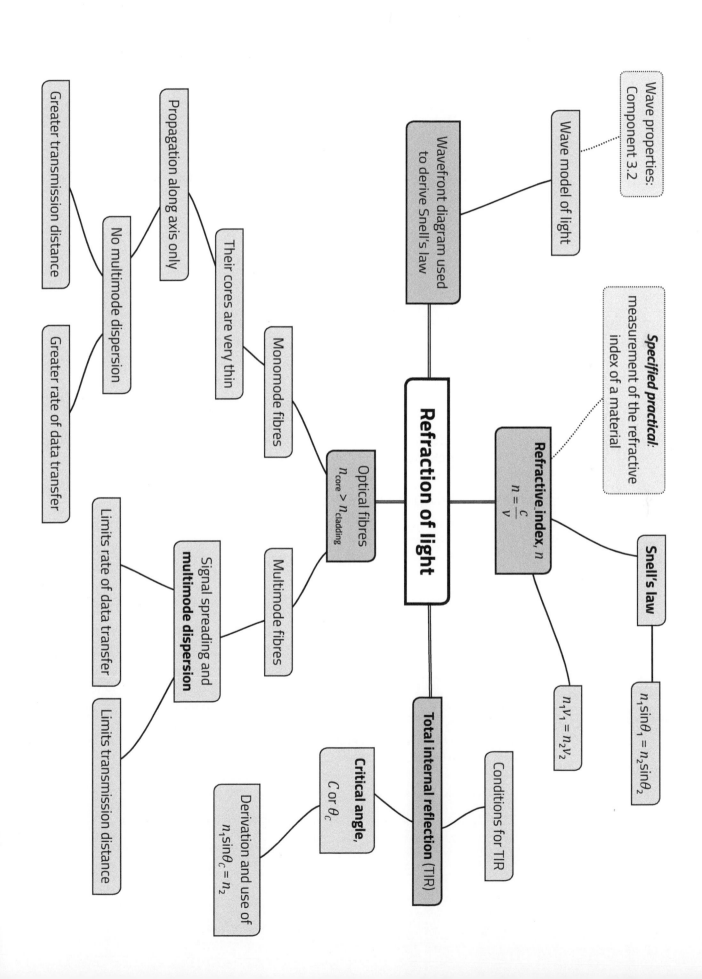

Q1 (a) State Snell's law. [2]

....

....

....

(b) Use Snell's law to define the term *refractive index* of a material in terms of angles. [2]

....

....

....

(c) Jennifer said that there was an equally good definition of *refractive index* in terms of the speed of light. Give this definition. [2]

....

....

....

Q2 Calculate the speed of light in a medium of refractive index 1.49. [2]

....

....

....

Q3 *Cherenkov radiation* is a form of radiation that can occur when electrons travel faster than the speed of light within a particular medium – it can be seen clearly in the cooling water surrounding nuclear reactors.

(a) State why Cherenkov radiation is impossible if the 'medium' is a vacuum. [1]

....

....

(b) (i) Calculate the minimum speed of electrons in water that will produce Cherenkov radiation ($n = 1.33$ for water). [2]

....

....

....

(ii) Calculate the pd required to accelerate an electron to this speed. Assume that the high velocity does not require Einstein's Theory of Relativity. [2]

....

....

....

....

Q4 A beam of light passes from air through a parallel-sided glass sheet (n = 1.52) and into the water (n = 1.33) of a fish tank, as shown. The initial angle of incidence is 45.0°. The angles are not drawn accurately.

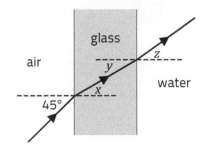

(a) Without calculation, explain why the light changes direction as shown at the two surfaces. [2]

....

....

....

(b) Calculate angles x, y and z. [3]

....

....

....

....

....

....

(c) Jordan says that you can calculate the angle z without first calculating y. Evaluate this statement without calculations. [2]

....

....

....

Q5 Light is incident upon a sphere of refractive index 1.42 as shown (the normal at the point of incidence has been added to the diagram).

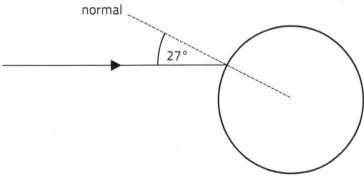

(a) Calculate the refracted angle as light enters the sphere **and draw it in the diagram**. [3]

....

....

....

(b) Draw two continuations of the ray to show the light being **reflected and refracted** at the point where the light ray you have drawn in part (a) is incident upon the opposite side of the sphere. [2]

Q6 Light in air is incident upon a glass prism of refractive index 1.55 at an angle, i, to the normal of the front surface. The prism's cross-section is an equilateral triangle (see below). The angles i and c are not drawn with their correct values.

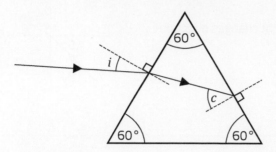

The light refracts at the front surface of the prism but when it is incident upon the second surface, it is incident at the critical angle, c, of the glass.

(a) State the conditions required for total internal reflection. [2]

....

....

....

(b) Calculate the critical angle of the glass. [2]

....

....

....

(c) Calculate the incident angle, i. [3]

....

....

....

....

....

(d) Briony claims that, because the light is incident on the second surface at the critical angle, the light will never leave the prism and will be totally internally reflected each time the ray is incident upon one of the prism's surfaces. Determine whether, or not, Briony is correct. [3]

....

....

....

....

....

Q7 Explain how you would carry out an experiment to measure the refractive index of a glass block. You should also refer to your **graphical** method of analysing your results. [6 QER]

....

....

....

....

....

....

....

....

....

....

....

....

Q8 A student designs an alarm system to detect when the level of benzene ($n = 1.51$) in a tank drops below the minimum height. The arrangement is shown in the diagram. The prism is fixed in position and made of glass with a refractive index of 1.50.

Explain how the system is meant to operate. [4]

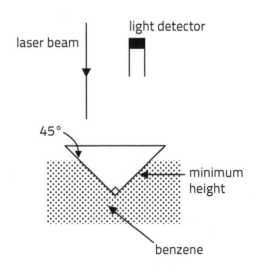

....

....

....

....

....

....

....

Q9 Light is incident upon an optical fibre as is shown in the diagram.

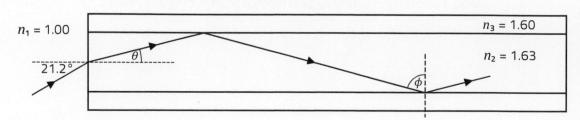

(a) Calculate the angles θ and ϕ. [3]

....

....

....

....

(b) Explain why light transmitted at this angle will not undergo total internal reflection. [3]

....

....

....

....

(c) The ray shown in the diagram below does undergo total internal reflection.

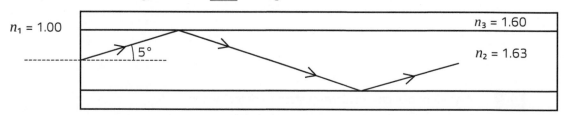

(i) Calculate the extra distance this light ray travels along a 14.0 km optical fibre compared with a light ray that travels straight along the axis of the fibre. [2]

....

....

....

(ii) Hence, explain the term *multi-mode dispersion*. [3]

....

....

....

....

....

Q10   An isosceles right-angled prism, **ABC**, is made from glass of refractive index 1.60. The face, **AC**, of the prism is horizontal. A narrow horizontal beam of light is incident from air on the face **AB** of the prism as shown in the diagram. The refracted beam is subsequently incident on face **AC**.

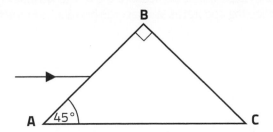

(a)   Show that the light is totally reflected at face **AC** and subsequently emerges from face **BC** parallel to its original direction. Add to the diagram to illustrate this.   [5]

(b)   A second light beam, parallel to the first, is incident on face **AB** nearer to **A** than the first beam. Add the path of this light ray to the diagram and suggest why such a prism can be used as an inverting prism (i.e. objects viewed through it appear upside down).   [3]

(c)   James said that, whatever the refractive index of the glass, the angle of incidence on face **AC** will always be greater than the critical angle. Hence it doesn't matter what the refractive index of the glass is, in this inverting prism arrangement. Discuss to what extent he is correct.   [4]

Question and mock answer analysis

Q&A 1 Geraint carries out an experiment to measure the refractive index of a glass block. He employs the following apparatus and records his results in the following table.

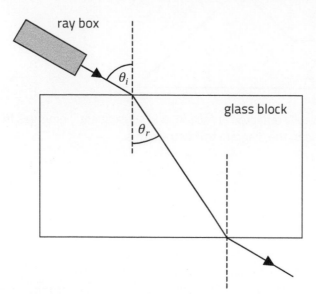

$\theta_i/°$	$\theta_r/°$	$\sin \theta_i$	$\sin \theta_r$
0	0	0	0
10	7	0.17	0.12
20	13	0.34	0.22
30	20	0.50	0.34
40	26	0.64	
50	31		0.52
60	36	0.87	0.59
70	39	0.94	0.63
80		0.98	0.66

(a) Complete the table. [3]

(b) Plot a graph of $\sin \theta_r$ (on the y axis) against $\sin \theta_i$ (on the x-axis), making best use of the grid, and draw a line of best fit. [Note: Page of graph grid provided, 12 × 9 squares] [5]

(c) Discuss to what extent the graph confirms Snell's law (that is, $\sin \theta_i \propto \sin \theta_r$). [3]

(d) Determine a value for the refractive index of the glass block. [3]

What is being asked

This is an experimental analysis question set around a specified practical. It is assumed that the candidates will have experience of undertaking it. Part (a) is set to get the candidates to start interacting with the data. Part (b) is a standard graph question with the added twist that there is a requirement to make best use of the grid, which is slightly more than the standard demand that the points occupy more than half of each scale. Part (c) is an evaluation which appears on Eduqas papers frequently and part (d) requires candidates to determine the relationship between the gradient of the graph and the refractive index.

Mark scheme

Question part	Description	AOs			Total	Skills	
		1	2	3		M	P
(a)	$40°$: $\sin\theta_r = 0.44$ **and** $50°$: $\sin\theta_i = 0.77$ [1] $80°$: $\theta_r = 41°$ [1] All to 2 sf [1]		3		3	3	3
(b)	Axes correctly orientated and both labelled [1] Landscape grid used with maximum scale [1] All points plotted within half a square (ignore 0) [2] (6 points plotted within half a square [1]) Line of best fit accurate – must be straight and, ideally, from the origin to slightly above the last point within half a square [1]	1	4		5		5
(c)	Straight line [1] Through the origin [1] Points are close to the line of best fit OR small scatter [1] ∴ Good agreement [needed for 3 marks]			3	3		3

| (d) | | Realising that 1/gradient is the refractive index [1]
 Correct method for calculating the gradient [1]
 Correct final answer (1.49, allow ± 0.02) [1]
 Alternative
 choosing values on the line of best fit (✓)
 Realising $n = \dfrac{\sin \theta_i}{\sin \theta_r}$ (✓)
 Correct final answer (1.49 ± 0.02) (✓)
 Max mark for choosing data point not on line = 2/3 marks | | | 3 | 3 | 3 | 3 |
| **Total** | | | 1 | 7 | 6 | 14 | 6 | 14 |

Rhodri's answers

(a) 0.44, 0.76 ✗
     41.3 ✓ ✗

MARKER NOTE
Rhodri has not rounded the second number correctly, so loses the first mark. He gains the second mark for the value but not the third because he does not express it to 2 sf.

1 mark

(b)

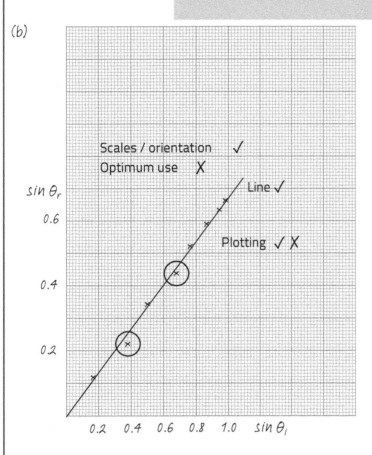

Scales / orientation ✓
Optimum use ✗
Line ✓
Plotting ✓ ✗

$\sin \theta_r$
0.6
0.4
0.2

0.2   0.4   0.6   0.8   1.0 $\sin \theta_i$

MARKER NOTE
Rhodri has labelled the axes correctly and they have the right orientation, so he gains the first mark.
To gain the second mark he should have used the grid landscape (as Ffion) as this makes fuller use of the grid.
He has plotted most of the points correctly but on the $\sin \theta_i$ axis has plotted 0.38 instead of 0.34 and 0.68 instead of 0.64 (circled points) so only gains one plotting mark.
The mark for the line is awarded because it is an acceptable line for the points as plotted.

3 marks

(c) Quite good agreement cos it's linear ✓
     but the data could be better cos the 2nd and
     4th points are a bit away from the line. ✓ (ecf)

MARKER NOTE
Rhodri's English leaves a little to be desired but 'linear' is equivalent to straight line and he merits the first mark. The 2nd cannot be awarded 'cos' he hasn't mentioned the origin. He obtains the 3rd mark because, with ecf, his comments regarding the 2nd and 4th data points are correct.

2 marks

(d)   Using $n = \dfrac{\sin i}{\sin r}$ ✓ $= \dfrac{0.34}{0.22}$ X

hence $n = 1.55$ X

MARKER NOTE
Rhodri has obtained a value by using data from the table. Unfortunately, he has chosen the point that is furthest from the line of best fit. This has led to a value of n outside the accepted limits and the only mark he can gain is the 2nd mark.

1 mark

| Total | 7 marks /14 |

Ffion's answers

(a) 0.44, 0.77 ✓
41 ✓✓

MARKER NOTE
Ffion's numbers are identical to those in the mark scheme, so full marks. Note: these are quite easy marks and one would expect most students to obtain full marks here.

3 marks

(b)

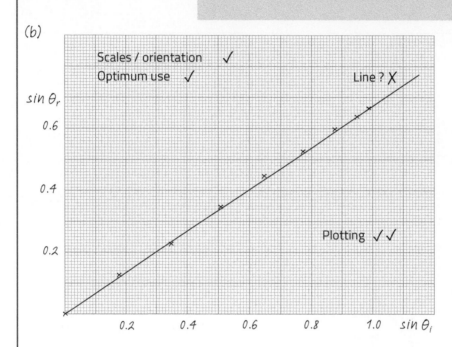

MARKER NOTE
Ffion's axes are labelled and have the correct orientation (1st mark). The graph is also the maximum size, without choosing difficult scales (2nd mark). Ffion's data points are all plotted correctly. She thus obtains the first 4 marks. The last mark is difficult to judge. Although Ffion's line is within half a square of being exact, she has 6 points above the line and only 1 point below it. The examiner has gone with X but will discuss it with the senior examiner.

4 or 5 marks

(c) It's a straight line with a positive gradient ✓ The scatter about the line of best fit is quite small and is to be expected considering that the angles are only measured to the nearest degree. ∴ Good agreement ✓

MARKER NOTE
The mark that Ffion misses is the second one – she needed to mention that the best-fit line passed through the origin. She has gone further and explained the existence of scatter but unfortunately there are no marks for this on the mark scheme.

2 marks

(d)   Refractive index is the gradient. ?

Gradient $= \dfrac{0.7}{1.045} = 0.766$ ✓

Hence, $n = \dfrac{1}{0.766} = 1.49$ ✓✓ (bod)

MARKER NOTE
Ffion's answer contains a contradiction. She starts by stating that the n = gradient which is incorrect. Had she stuck to this she would only have received one mark for calculating the gradient. However, she later does 1/gradient to obtain a final answer that is correct. Although there is a mistake in the first line which has not been crossed out, it is clear that she has corrected this by obtaining the final correct value. Ffion obtains the benefit of the doubt here because her final answer is correct.

3 marks

| Total | 12 or 13 marks /14 |

Section 4: Photons

Topic summary

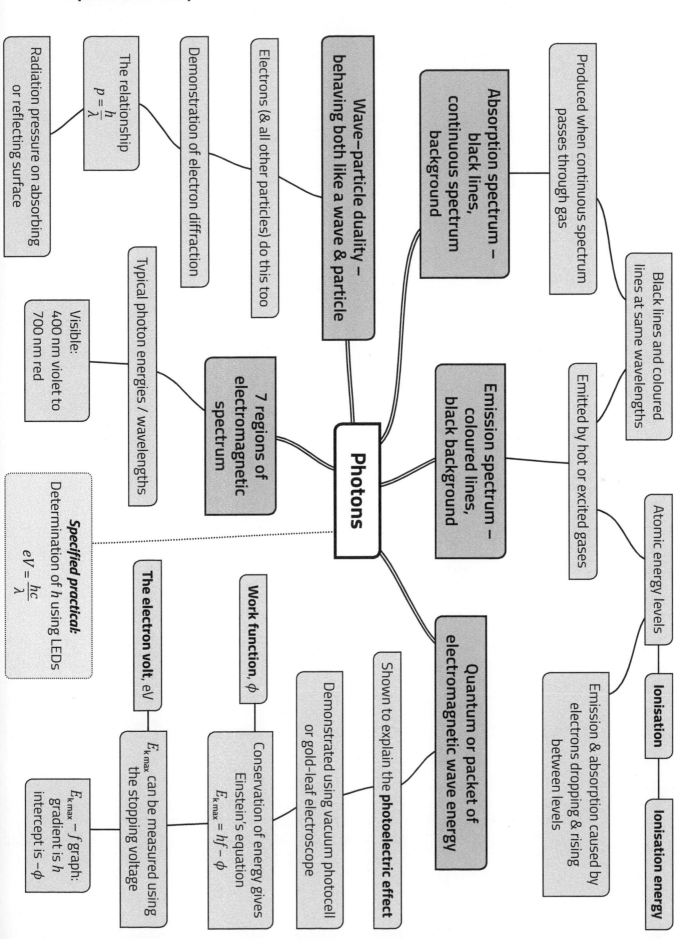

Radiation pressure on absorbing or reflecting surface

The relationship
$$p = \frac{h}{\lambda}$$

Demonstration of electron diffraction

Electrons (& all other particles) do this too

Wave–particle duality – behaving both like a wave & particle

Absorption spectrum – black lines, continuous spectrum background

Produced when continuous spectrum passes through gas

Black lines and coloured lines at same wavelengths

Visible:
400 nm violet to 700 nm red

Typical photon energies / wavelengths

7 regions of electromagnetic spectrum

Photons

Emission spectrum – coloured lines, black background

Emitted by hot or excited gases

Atomic energy levels

Specified practical:
Determination of h using LEDs
$$eV = \frac{hc}{\lambda}$$

The electron volt, eV

Work function, ϕ

Demonstrated using vacuum photocell or gold-leaf electroscope

Shown to explain the **photoelectric effect**

Quantum or packet of electromagnetic wave energy

Emission & absorption caused by electrons dropping & rising between levels

Ionisation

Ionisation energy

$E_{k\,max}$ can be measured using the stopping voltage

Conservation of energy gives Einstein's equation
$$E_{k\,max} = hf - \phi$$

$E_{k\,max} - f$ graph:
gradient is h
intercept is $-\phi$

Q1 A monochromatic beam of light has frequency, f, and power, P. It can also be described in terms of *photons*.

(a) State briefly what is meant by a photon. [1]

....

(b) Show how the power of the beam (i.e. the energy transferred per second) relates to the frequency and number of photons per second. [2]

....

....

Q2 Describe the appearance of a line emission spectrum, the physical situation which can give rise to it and how it can be displayed. [4]

....

....

....

....

....

....

Q3 Explain how an absorption spectrum is produced in the atmosphere of the Sun. [4]

....

....

....

....

....

....

Q4 (a) Calculate the de Broglie wavelength of an electron accelerated by a pd of 2400 V. [4]

....

....

....

....

....

....

(b) An electron diffraction experiment produces a diffraction pattern of concentric rings. Explain what happens to this pattern as the accelerating pd is increased. [2]

....

....

....

Q5 The following diagram shows some of the energy levels of atomic hydrogen.

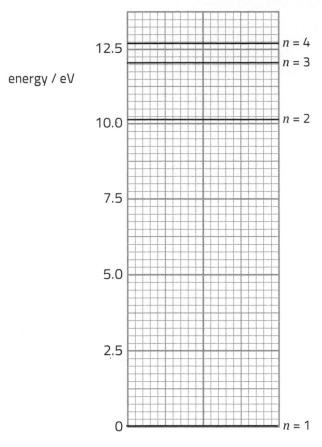

(a) (i) Calculate the energy of a photon, in eV, that is emitted when an electron drops from energy level $n = 3$ to energy level $n = 2$. Give a reason for your answer. [2]

....

....

....

(ii) Calculate the wavelength of this photon and state in which region of the electromagnetic spectrum it belongs. [2]

....

....

....

(b) (i) Calculate the highest energy, in J, of a photon that can be absorbed in a transition between the energy levels shown in the diagram. [2]

....

....

....

(ii) State in which region of the electromagnetic spectrum this photon resides. [1]

....

Q6 (a) Explain Einstein's photoelectric equation, its importance historically and how the apparatus below can be used to obtain the graph of results (also shown). [6QER]

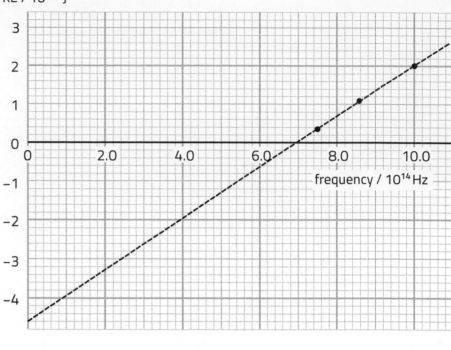

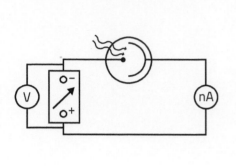

....

....

....

....

....

....

....

....

....

....

....

....

(b) Explain why the current produced by the photoelectric effect in the apparatus in part (a) is proportional to the light intensity, for a given frequency of light. [The pd of the supply has been set to zero.] [3]

....

....

....

....

Q7 The use of lasers to drive unpowered spacecraft has been the subject of research by scientists and speculation by science fiction writers. This question looks at the principles involved. Gravitational forces can be neglected in your answers.

A space probe without a rocket engine is sent to Mars. It is launched from the International Space Station and accelerated using a 15 kW laser which is located on Earth. The thrust on the probe is produced by a mirror which reflects the laser light incident on the probe.

(a) Explain briefly why reflecting the laser light can produce a force on the probe. [2]

....

....

....

(b) (i) Show that the momentum, p, of a photon is related to its energy, E, by the relationship:

$$E = pc$$

where c is the speed of light. [2]

....

....

....

(ii) The mass of the probe is 2.3 kg and all the laser light is reflected directly back continuously by the probe. Calculate the acceleration of the probe (it may be beneficial to use the relationship in (b)(i)). [3]

....

....

....

....

....

(iii) After 1 year, calculate:

(I) the speed of the probe, [2]

....

....

....

(II) the distance travelled by the probe. [2]

....

....

....

Component 3 Practice questions

(iv) A mirror was placed on Mars by a previous mission. Explain how this might be used to decelerate the probe. [2]

(c) One problem with using a laser in this way arises from the fact that light has wave properties as well as particle properties. The laser beam emerges from a circular window, so diffraction occurs and the beam diverges as shown in the diagram:

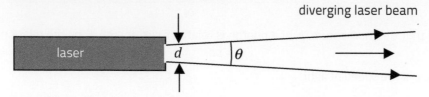

For a circular aperture, the angle of divergence, θ, is given, in radians, by:

$$\theta = \frac{2\lambda}{d}$$

where d is the diameter of the window.

(i) A lecturer uses a green laser pointer with a wavelength of 500 nm in a lecture theatre of length 10 m. Using a sensible estimate of d show that diffraction of the laser beam is unlikely to be a problem. [4]

(ii) In an attempt to make the green space laser as effective as possible it has a window of diameter 1 m. The spacecraft has a receiving surface of diameter 100 m. Estimate how far the spacecraft travels before the driving force is 10% of the maximum. Comment on your answer. [4]

Question and mock answer analysis

Q&A 1   The following circuit is used to find the pd across a LED when it is switched on:

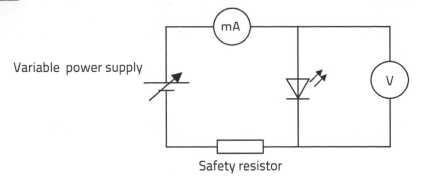

The LED is judged to be switched on when a current of 10.0 mA passes through it. The variable power supply is adjusted and the 'switching on pd', V_s, recorded when the current reaches 10.0 mA. The procedure is repeated for a range of different LEDs which emit light of different wavelengths, λ. The incomplete results and plot of them are shown below:

λ / nm	$\frac{1}{\lambda}$ / 10^6 m^{-1}	V_s / V
480	2.08	2.65
590		2.14
680	1.47	1.85
895		1.40
930	1.08	1.34

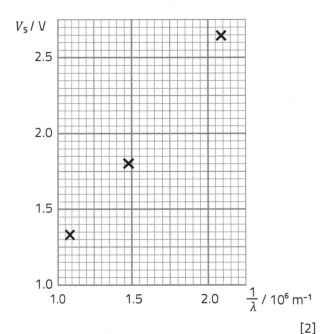

(a)   Complete the table.   [2]

(b)   Complete the graph by plotting the two missing points whose values you have just calculated and drawing an appropriate line.   [3]

(c)   The application of the principle of conservation of energy to an electron and a photon involved in the light-emitting process of the LED gives:
$$eV = \frac{hc}{\lambda}$$

(i)   Use the graph to determine a value for the Planck constant to an appropriate number of significant figures.   [4]

(ii)   Explain to what extent the data confirm the relationship, $eV = \frac{hc}{\lambda}$.   [5]

What is being asked?

This is a specified practical, with which you should be familiar. However, there is no need to recall any aspect of the practical work as all the questions involve interaction with the data and forming conclusions, i.e. the entire question is AO2 and AO3. The question style is quite a familiar one in the A Level examination. Students can be asked to perform calculations on experimental data, measure gradients and intercepts and draw conclusions about whether the data agree with theoretical ideas.

Mark scheme

Question part			Description	AOs			Total	Skills	
				1	2	3		M	P
(a)			1.69 [1] 1.12 [1]		2		2	2	2
(b)			(1.69, 2.14) plotted to within ½ square in each direction – ecf [1] (1.12, 1.40) plotted to within ½ square in each direction – ecf [1] Good fit straight line drawn (by eye) to plotted points [1]			3	3		3
(c)	(i)		Correct method for calculating gradient, i.e. $\Delta y / \Delta x$ attempted. [1] Correct gradient (ecf on line) – expect 1.3 [× 10⁶] – ignore powers of 10. [1] Statement that gradient = $\frac{hc}{e}$ [or by implication] [1] Correct answer with 2 or 3 sf only (expect 6.9[0] × 10⁻³⁴) [1] [If data point from the line or table is used, max 2 marks allowed if 2 or 3 sf given]			4	4	4	4
	(ii)		[The line of best fit] is a straight line [1] Intercept on V_s calculated [1] Comment that intercept is very close to origin [1] Very little scatter in data points **or** points close to line of best fit. [1] The value of the Planck constant is close to the accepted [accept: actual] value. [1]			5	5	1	5
Total				0	5	9	14	7	14

Rhodri's answers

(a) 1.69 ✓

   1.1173 ✗ [rounding]

MARKER NOTE
Rhodri has calculated $1/\lambda$ correctly. He should have rounded up the second answer [$1.1173 \rightarrow 1.12$ to 2 sf]. He gets the mark for the first answer.

1 mark

(b)

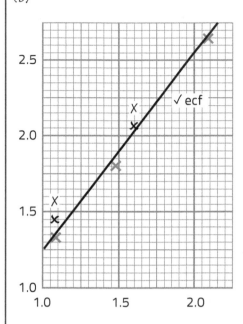

MARKER NOTE
Rhodri has mis-plotted both his data points. He has misinterpreted the scale, thinking that each small square is 0.1 on both axes, whereas in fact two squares are 0.1. So no plotting marks. The poor plotting makes it more difficult to judge the best-fit line but it is a reasonable attempt and gains the line mark.

1 mark

(c) (i) Using the middle point: $\lambda = 680$ nm and $V_s = 1.85$ ✗
   Rearrange equation: $h = \dfrac{eV\lambda}{c}$ ✓

   $h = \dfrac{1.6 \times 10^{-19} \times 1.85 \times 10^{-9}}{3 \times 10^{8}}$ ✗

   $h = 6.709 \times 10^{-34}$ ✗

MARKER NOTE
Rhodri has obtained a good value for the Planck constant but he has used a single data point rather than the gradient. This is especially wrong because the line doesn't quite pass through the origin, possible because of a systematic error. He should have worked to a maximum of 3 sf because the data are all to 3 sf, but he has given his answer to 4 sf. A 2 or 3 sf answer, i.e. 6.7 or 6.70, would have gained him two marks but unfortunately he loses one of these and only obtains one.

1 mark

(ii) The graph is a straight line ✓
   with a positive gradient but it doesn't
   pass through the origin. ✗ My value for
   Planck's constant is good. ✓

MARKER NOTE
Rhodri's evaluation is extremely succinct but, nonetheless, it hits two of the marking points: the straight line and the Planck constant value. He has not calculated the intercept on the V_s axis, so really has no idea whether it is zero or not – he has probably not noticed that neither axis scale goes back to zero. No additional credit is given for the positive gradient.

2 marks

Total — **5 marks /14**

Ffion's answers

(a) 1.70 ✗ [rounding]

    1.12 ✓

MARKER NOTE

Ffion has calculated the values and rounded the second one correctly. She has probably rounded the first answer twice in her head, a common mistake, i.e. 1.6949 → 1.695 → 1.70 which is incorrect rounding, so loses the first mark.

1 mark

(b)

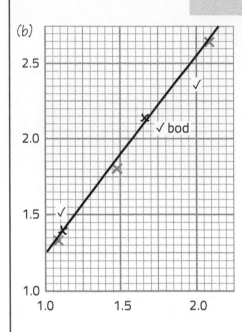

MARKER NOTE

Ffion's lower point is plotted in exactly the right place (1.12, 1.40). The upper point is a little bit out – it looks like (1.67, 2.13). This is just about within half a square in each direction from the real data point, so she is given this mark with a tiny bod.

Because of her accurate plotting, the best fit line is quite straightforward to draw and she picks up the line mark.

3 marks

(c) (i) Gradient = $\dfrac{2.75 - 1.23}{(2.17 - 1.00) \times 10^6}$ ✓

          = 1.299 × 10⁻⁶ ✓

      But gradient = $\dfrac{hc}{e}$ ✓

      So h = $\dfrac{e}{c}$ × gradient = 6.93 × 10⁻³⁴ J s ✓

MARKER NOTE

Ffion's answer is perfect in every way. Note that she keeps her intermediate value of gradient to 4 sf in case of rounding errors. Her final answer to 3 sf is correct even though, at first glance, Rhodri's appears more accurate. Another particularly pleasing thing about Ffion's answer is the figures she provides in calculating the gradient. Pay particular attention to the value of the y-axis intercept – this is good precision.

4 marks

   (ii) The line of best fit is a staight line ✓ which passes through all the data points ✓

      When the x value is 0 the y value is :

      1.23 − (1.0 × 10⁶) × 1.30 × 10⁻⁶ = −0.07. ✓

      Theory predicts the graph should go through zero, so this is not quite right but quite close. ✓

      The value obtained for the Planck constant is close to the accepted value of 6.63 × 10⁻³⁴ Js (only 4% out). ✓

MARKER NOTE

Ffion's answer mirrors the mark scheme and is, in fact, better than the mark scheme (which tends to state the minimum that is required to obtain the marks). The one criticism is that her explanation of how she calculated the y-intercept could have been clearer, but it was good enough for the examiner to follow her reasoning.

5 marks

| **Total** | **13 marks /14** |

Section 5: Lasers

Topic summary

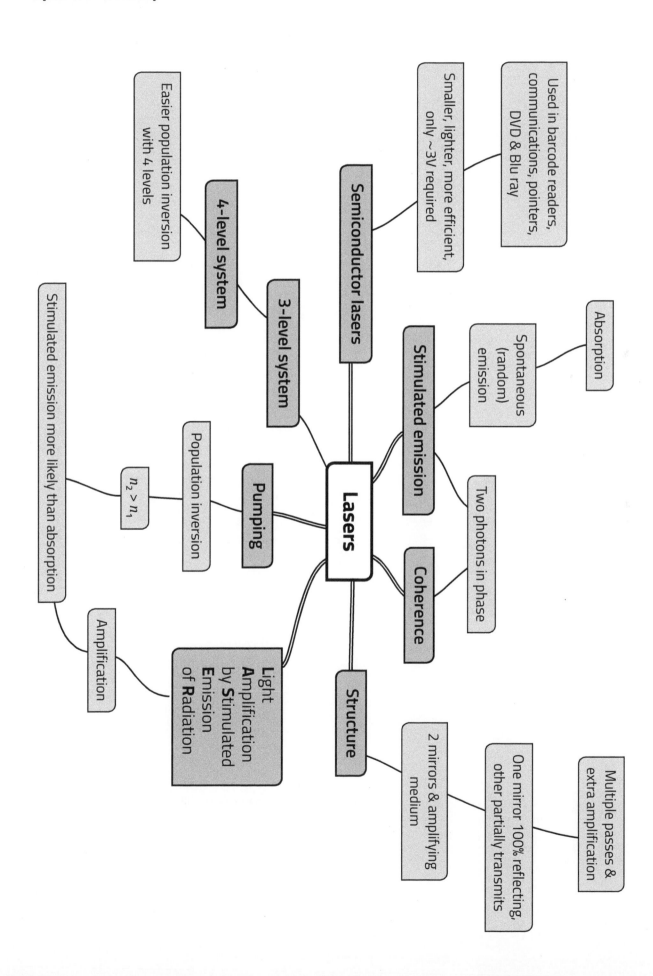

Easier population inversion with 4 levels

4-level system

3-level system

Semiconductor lasers

Smaller, lighter, more efficient, only ~3V required

Used in barcode readers, communications, pointers, DVD & Blu ray

Stimulated emission more likely than absorption

$n_2 > n_1$

Population inversion

Pumping

Amplification

Light Amplification by Stimulated Emission of Radiation

Lasers

Stimulated emission

Spontaneous (random) emission

Absorption

Two photons in phase

Coherence

Structure

2 mirrors & amplifying medium

One mirror 100% reflecting, other partially transmits

Multiple passes & extra amplification

Q1 Laser is an acronym for Light Amplification by Stimulated Emission of Radiation. Explain what is meant by the term *stimulated emission*. [3]

....

....

....

....

....

Q2 (a) State what is meant by the term *population inversion*. [1]

....

....

(b) Explain why it is not possible to obtain a *population inversion* in a 2-level system when optical pumping is used. [4]

....

....

....

....

....

....

Q3 Explain why a population inversion is harder to obtain in a 3-level laser system than a 4-level laser system. You should add to the diagrams as part of your answer. [4]

3-level   4-level

....

....

....

....

Q4 The pumping level (top level) in laser systems must have a short lifetime. State two reasons why the pumping level must have a short lifetime. [2]

....

....

....

Q5 State two uses and two advantages of semiconductor lasers. [2]

....

....

....

Q6 The energy levels of a 4-level laser system are shown:

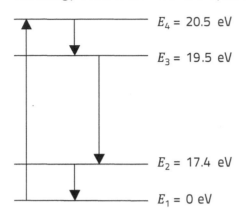

$E_4 = 20.5$ eV

$E_3 = 19.5$ eV

$E_2 = 17.4$ eV

$E_1 = 0$ eV

(a) Calculate the frequency of laser emission. [3]

....

....

....

....

....

(b) Laser emission occurs when an electron is stimulated to drop from the 3rd energy level. State what is responsible for stimulating this emission and explain why stimulated emission is far more probable than spontaneous emission for a laser. [3]

....

....

....

....

(c) Joel claims that this particular laser system could never be more than 10% efficient. Evaluate to what extent Joel is correct. [2]

....

....

....

(d) Nigella states, 'The 2nd energy level (E_2) in a 4-level laser system must have a long lifetime.' Evaluate to what extent Nigella's statement is correct. [2]

....

....

....

Q7 The energy levels of a 3-level laser system are shown.

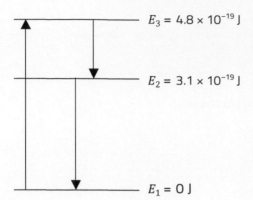

$E_3 = 4.8 \times 10^{-19}$ J

$E_2 = 3.1 \times 10^{-19}$ J

$E_1 = 0$ J

(a) Calculate the energy of the photons used to pump the laser in eV and state what colour they are. [2]

....

....

....

(b) Compare the lifetimes of the two excited states in this system and give a reason for your answer. [2]

....

....

....

(c) Calculate the wavelength of laser emission. [2]

....

....

....

(d) Two photons travel together after stimulated emission occurs. State three distinct properties that are common to these two photons. [3]

....

....

....

(e) Paula states 'More than half the electrons of the ground state must be pumped in order to achieve population inversion in a 3-level laser system.' Evaluate to what extent Paula's statement is correct. [3]

....

....

....

....

....

Q8 A diagram of a laser cavity is shown below:

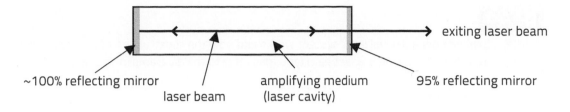

~100% reflecting mirror   amplifying medium   95% reflecting mirror
        laser beam   (laser cavity)   exiting laser beam

(a) In a laser, the left mirror cannot be made to be exactly 100% reflecting. Explain why the left mirror should, ideally, be 100% reflecting. [2]

....

....

....

(b) Explain why the right mirror must not be 100% reflecting. [1]

....

....

(c) (i) Explain briefly why the force exerted by the laser light on the left mirror is greater than the force exerted on the right mirror. [2]

....

....

....

(ii) The output power of the laser is 5 mW and the cross-sectional area of the laser beam is 0.94 mm². Calculate the pressure exerted on the right mirror (you may assume that the wavelength of the laser light is 500 nm although this problem is soluble without this information). [3]

....

....

....

....

(d) Helena claims that the laser must increase in intensity by approximately 3% each time it traverses a length of the laser cavity (when the laser is in equilibrium). Evaluate whether or not Helena's claim is correct. [3]

....

....

....

....

Question and mock answer analysis

Q&A 1

(a) Explain how a 3-level laser system works **and** how the construction of the laser cavity and mirrors assists in producing a laser beam (you may refer to the diagrams in your answer). [6 QER]

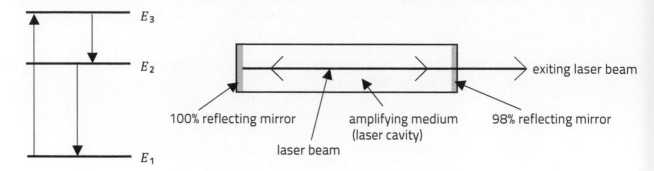

(b) Light of wavelength 950 nm from a powerful 50 W laser is focused from a beam diameter of 1.2 mm to a beam diameter equal to one wavelength (950 nm). Calculate the final intensity of the laser beam. [3]

What is being asked

(a) The specification has a series of statements about laser function, all of which are addressed in this question. The statements are:
- the process of stimulated emission...
- the idea that a population inversion ($N_2 > N_1$) is necessary for a laser to operate
- how a population inversion is attained in 3-level energy systems
- the process of pumping and its purpose
- the structure of a typical laser, i.e. an amplifying medium between two mirrors, one of which partially transmits light.

This question asks you to pull together these statements in order to write an account of laser operation. This is a very large topic but the examiner has used the diagrams to provide a framework for this AO1 piece of bookwork. Notice that there are two broad parts to the question. You are expected to tackle both – hence the emboldened **and** in the question.

The marking scheme has a series of points which you could choose to make. You are not expected to mention all of them but you need to use several of them from both sections (energy levels and structure) in order to be put into the top band.

(b) The second part of the question is an AO2 calculation and will require the application of conservation of energy and an understanding of the term intensity. This question is synoptic because intensity normally appears with the topic of stars in Component 1. It has an added level of difficulty because the definition of intensity in the Terms & Definitions booklet does not quite cover this question – that definition is aimed at stars and the inverse square law.

Mark scheme

Question part			Description	AOs			Total	Skills	
				1	2	3		M	P
(a)			**Points regarding energy levels**	6			6		
			1. Pumping is from E_1 to E_3						
			2. E_3 has a short lifetime						
			3. E_2 is metastable / has a long lifetime						
			4. E_1 is the ground state						
			5. Ground state is usually full						
			6. Population inversion between E_2 and E_1						
			7. Stimulated emission between E_2 and E_1						
			8. Greater than 50% pumping needed						
			Points regarding mirrors and structure						
			1. Amplification while passing through cavity (exponential growth)						
			2. 100% mirror returns beam to cavity						
			3. 98% mirror allows 2% through (but returns/reflects nearly all)						
			4. Multiple passes (~50) through cavity						
			5. Equilibrium when overall losses through the 98% mirror = overall gains due to amplification						
			6. Parallel mirrors give alignment of beam						
			7. Stationary wave between mirrors						
(b)			Realising power will still be 50 W [1]		3		3	1	
			Application of $I = \dfrac{P}{\lambda r^2}$ or $I = \dfrac{P}{\text{area}}$ [1]						
			Correct answer – 70.5 TW m^{-2} [1]					1	
Total				**6**	**3**		**9**	**2**	

Rhodri's answers

(a) The transition from E_1 to E_3 is called pumping. E_3 is a broad energy level with a very short lifetime to help with pumping and keeping E_3 empty. E_2 has a long lifetime so that a population inversion can be obtained between E_2 and E_1. It is difficult to obtain a population inversion because E_1 is the ground state and is normally full. Hence, more than half the electrons must be pumped in order to obtain a population inversion. Stimulated emission will then be more likely than absorption between E_2 and E_1 and the laser beam will gain in intensity.

MARKER NOTE
Rhodri's answer is an excellent response to the first part of the question. However, he has forgotten to answer the second part of the question even though the 'and' has been emboldened in the question. This happens surprisingly often even with excellent candidates. Note that the marker has underlined Rhodri's points and that Rhodri has succeeded in hitting all the 8 energy level points in the MS. Rhodri has gone even further and has provided three extra pieces of information. First, E_3 is a broad level – true this helps to increase the probability of pumping. Second, he has correctly pointed out that E_3 should remain empty. Third, Rhodri has stated that stimulated emission is more probable than absorption, an excellent point not on the mark scheme. All this means that Rhodri has a middle band answer – an excellent response to only one of the two answer parts. However, the quality and clarity are such that 4 marks are deserved rather than 3 marks.

4 marks

(b) The intensity will still be 50 W ✗ because you won't lose any of the power of the laser. ✓

MARKER NOTE
Rhodri appears to have no concept of intensity. Nonetheless, he gains one mark because he has applied his knowledge of conservation of energy well.

1 mark

Total	5 marks /9

Ffion's answers

(a) The rise from E_1 to E_3 is called pumping and is often done by accelerated electrons or light absorption. E_3 must have a short lifetime so that the electrons go to E_2 quickly. Metastable level is the name for energy level E_2 and it has a long half-life. This means that a population inversion is possible between E_2 and E_1 leading to lots of stimulated emission and amplification. Amplification by traversing the laser cavity once is not usually enough and this is where the design of the laser system and mirrors is important. One mirror is 100% reflecting and this ensures that the beam returns to traverse the amplifying medium again, leading to greater amplification. The other mirror allows 2% of the light to be transmitted and this is the output beam. Only 2% of the light is output each time which means that photons traverse the amplifying medium 50 times on average before exiting.

(b) Intensity $= \dfrac{power}{area} = \dfrac{50 \ W}{4\pi r^2}$ ✓ [application]

$= \dfrac{50 \ W}{4\pi \left(950 \times 10^{-9} \ m\right)^2}$ ✓ ✗

$= 4.4 \times 10^{12} \ W \ m^{-2}$

MARKER NOTE

Ffion's answer has 4½ good points from the energy levels part. The marker has underlined the 4 points that are not in doubt. Ffion has also mentioned stimulated emission but has not stated explicitly between which levels this occurs although the stimulated emission has been linked to the population inversion. Also, Ffion has added a little extra explanation that the short half-life of E_3 allows electrons to pass quickly to E_2. Ffion's answer to the second part of the question also makes four good points (the first four points). Although Ffion has not hit any of the last 3 points on the 'mirrors & structure' part, these are more advanced points. Overall, Ffion's response to both parts is just about good enough to place her in the top band. Her use of English is good and the sentence linking both parts of the answer is a particularly nice touch. Final mark – 5 or 6 depending on whether the marker feels that rewarding the well-constructed answer is more important than penalising the omission of reference to the ground state.

5 or 6 marks

MARKER NOTE

Ffion does not mention that the power of the beam will still be 50 W but it is implied in her calculation and she receives this first mark. She realises that the initial area of the beam and the wavelength are unnecessary pieces of information.

She has attempted power / area to obtain the intensity and thus gained the second mark, but has made two separate mistakes. First, she has used the wrong equation for the area – she needs the cross-section of a circle not a sphere. Second, she has used the diameter of the beam instead of the radius of the beam. She is probably a little fortunate to receive 2 marks but this type of answer would have been discussed in the examiners' meeting and this mark would have been agreed at that stage.

2 marks

| **Total** | **7 or 8 marks /9** |

Section 6: Nuclear decay

Topic summary

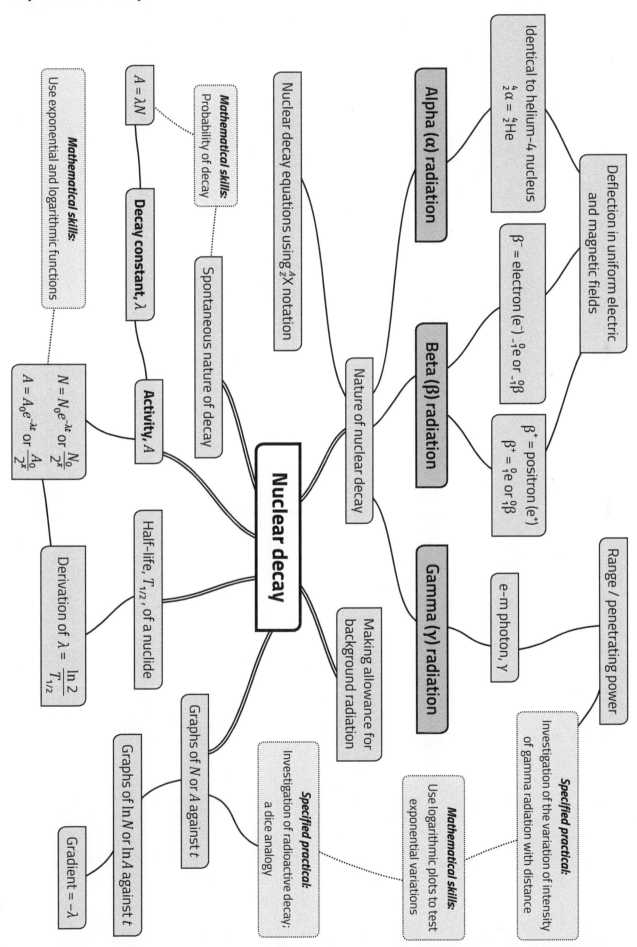

Nuclear decay

Mathematical skills:
Use exponential and logarithmic functions

$A = \lambda N$

Mathematical skills:
Probability of decay

Decay constant, λ

Nuclear decay equations using $_Z^A X$ notation

Spontaneous nature of decay

$N = N_0 e^{-\lambda t}$ or $\dfrac{N_0}{2^x}$

$A = A_0 e^{-\lambda t}$ or $\dfrac{A_0}{2^x}$

Activity, A

Derivation of $\lambda = \dfrac{\ln 2}{T_{1/2}}$

Half-life, $T_{1/2}$, of a nuclide

Graphs of N or A against t

Graphs of $\ln N$ or $\ln A$ against t

Gradient $= -\lambda$

Specified practical:
Investigation of radioactive decay;
a dice analogy

Making allowance for background radiation

Alpha (α) radiation

Identical to helium-4 nucleus
$_2^4 \alpha = _2^4 He$

Deflection in uniform electric and magnetic fields

Beta (β) radiation

$\beta^- = $ electron (e^-) $_{-1}^0 e$ or $_{-1}^0 \beta$

$\beta^+ = $ positron (e^+)
$\beta^+ = _1^0 e$ or $_1^0 \beta$

Nature of nuclear decay

Gamma (γ) radiation

e–m photon, γ

Range / penetrating power

Specified practical:
Investigation of the variation of intensity of gamma radiation with distance

Mathematical skills:
Use logarithmic plots to test exponential variations

Q1 When asked to describe what β radiation was, Alex wrote, 'It is a stream of electrons.' Charlie said that Alex's description was incomplete.

State what is missing from her description. [2]

....

....

....

Q2 The **activity**, A, of a radioactive sample can be calculated using the equation:

$$A = \lambda N$$

(a) State what is meant by the *activity* and give its unit. [2]

....

....

....

(b) The decay constant of plutonium-239 is 9.11×10^{-13} s^{-1}. A student reads that, in any year, an atom of plutonium-239 has a probability of less than 1 in 30 000 of decaying.

Evaluate whether this is correct. [3]

....

....

....

....

....

Q3 In an experiment to investigate the radiation emitted by a radioactive sample, a student set it up 10 cm from a radiation detector. Counts were taken over 5-minute periods with no absorber, a thin paper absorber and a 3 mm aluminium absorber. The results were as follows:

Absorber	none	paper	aluminium
Count	576	570	568

(a) Explain what conclusions can be drawn from the results about the emission of α, β and γ radiation by the sample. [3]

....

....

....

....

....

(b) Suggest two improvements to the experiment and explain how they will allow more complete conclusions to be drawn. [3]

....

....

....

....

Q4 Uranium-235 ($^{235}_{92}U$), decays by α-emission to an isotope of thorium (Th).

(a) Write the nuclear decay equation for $^{235}_{92}U$. [3]

$$^{235}_{92}U \longrightarrow$$

(b) The thorium isotope subsequently undergoes a series of α and β⁻ decays until a stable isotope of lead is formed. There are three stable isotopes of lead, $^{206}_{82}Pb$, $^{207}_{82}Pb$ and $^{208}_{82}Pb$.

(i) Paul says that the isotope formed must be $^{207}_{82}Pb$. Explain why he is correct. [2]

(ii) Determine the number of α decays and β⁻ decays that take place in the decay series from $^{238}_{92}U$ to $^{206}_{82}Pb$ and explain your answer. It might help you to complete the following equation (you may ignore neutrinos): [3]

$$^{238}_{92}U \longrightarrow {}^{206}_{82}Pb + \text{\_\_\_\_\_} + \text{\_\_\_\_\_}$$

(iii) The uranium isotope $^{233}_{92}U$ is part of a different radioactive decay series. Explain why this series cannot **end** on an isotope of lead. [2]

Q5 A teacher sets up a cobalt-60 gamma source 20 cm from a radiation detector and rate meter. The reading, corrected for background, is 9.76 counts per second. Exactly one year ago, the reading was 11.50 counts per second.

(a) Calculate what the reading will be:

(i) one year from now. [2]

(ii) in 10 years' time. [1]

(b) The level of background radiation in the teacher's lab is 0.42 counts per second. Determine the time it will take for the cobalt-60 source to decay to the same level as this measured at 20 cm distance. [3]

Q6 High speed particles from space, known as cosmic rays, collide with atoms in the upper atmosphere. Some of the collisions result in the emission of neutrons which are absorbed by nitrogen, $^{14}_{7}N$, nuclei, producing $^{14}_{6}C$ nuclei, which are β^- radioactive with a half-life of 5730 years. The balance, between the production of $^{14}_{6}C$ and its decay, results in a ratio of $^{14}_{6}C / ^{12}_{6}C$ atoms of 1.250×10^{-12}.

The tissues of living organisms contain $^{14}_{6}C$ in the same ratio to $^{12}_{6}C$ as in the atmosphere. After an organism dies, the level of $^{14}_{6}C$ in its tissues decreases, due to radioactive decay. This decrease can be used to estimate the age of objects made from biological materials in a process called radio-carbon dating.

(a) Complete the nuclear equation for the production of $^{14}_{6}C$. [2]

$$^{-}_{-}n + {}^{14}_{7}N \longrightarrow {}^{14}_{6}C +$$

(b) Write the nuclear decay equation for $^{14}_{6}C$. [2]

$$^{14}_{6}C \longrightarrow \text{...............} + \text{...............} + \text{...............}$$

(c) (i) Calculate the radioactive decay constant, λ, for carbon-14. [2]

....

....

....

(ii) A wooden artefact from ancient Egypt was found to have a $^{14}_{6}C / ^{12}_{6}C$ ratio of $(0.851 \pm 0.002) \times 10^{-12}$. Estimate the age of the wood together with its absolute uncertainty. [3]

....

....

....

....

....

(iii) The burning of fossil fuels in the last 200 years has reduced the amount of $^{14}_{6}C$ in the atmosphere by 3%, compared to $^{12}_{6}C$. Sioned says that this will make recently manufactured objects seem older than they really are, if assessed using radio-carbon dating. Evaluate whether Sioned is correct. [2]

....

....

....

....

....

Q7   A class of students investigates radioactive decay theoretically, by imagining an experiment involving 800 octahedral dice, each of which has one face painted black.

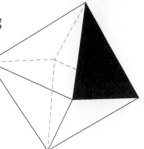

The students consider throwing the dice several times and, each time, removing any that land with the black face up.

(a) Show that the number of dice the students expect to remain after n throws is $800 \times (0.875)^n$.   [2]

....

....

....

(b) The students then perform the experiment with the following results:

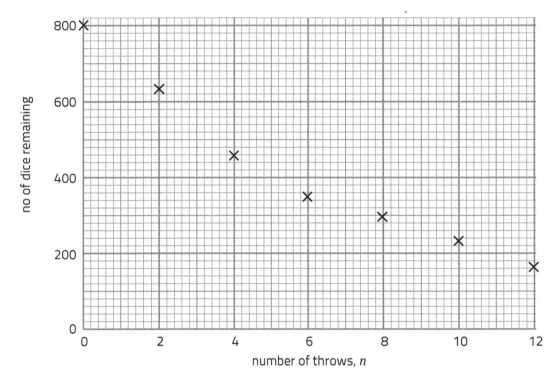

Comment on whether the results agree with the expected 'half-life' of the dice.   [3]

....

....

....

....

....

(c) The students paint a second face black on each of the dice and repeat the experiment. Draw the expected decay graph on the grid above.   [3]

Q8 β particles lose kinetic energy when they pass through the atoms of an absorber. They do this by interactions with the electrons in the atoms.

Their range in the absorber is the distance they travel until they lose all their kinetic energy.

The graph shows how the range of β particles **in water** depends upon their energy.

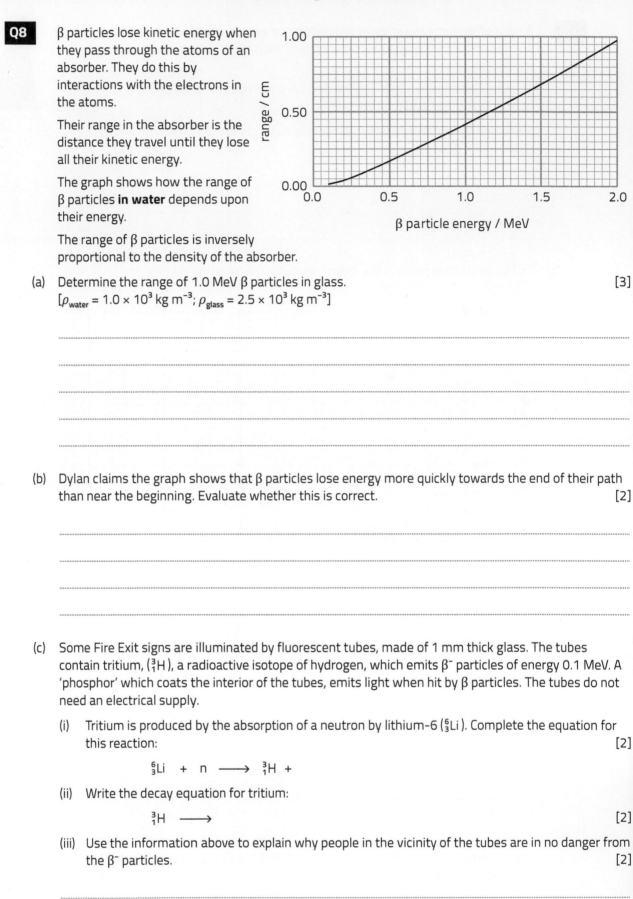

The range of β particles is inversely proportional to the density of the absorber.

(a) Determine the range of 1.0 MeV β particles in glass. [3]
$[\rho_{water} = 1.0 \times 10^3$ kg m^{-3}; $\rho_{glass} = 2.5 \times 10^3$ kg m^{-3}]

....

....

....

....

(b) Dylan claims the graph shows that β particles lose energy more quickly towards the end of their path than near the beginning. Evaluate whether this is correct. [2]

....

....

....

....

(c) Some Fire Exit signs are illuminated by fluorescent tubes, made of 1 mm thick glass. The tubes contain tritium, ($^3_1$H), a radioactive isotope of hydrogen, which emits β$^-$ particles of energy 0.1 MeV. A 'phosphor' which coats the interior of the tubes, emits light when hit by β particles. The tubes do not need an electrical supply.

(i) Tritium is produced by the absorption of a neutron by lithium-6 ($^6_3$Li). Complete the equation for this reaction: [2]

$$^6_3\text{Li} \; + \; n \; \longrightarrow \; ^3_1\text{H} \; +$$

(ii) Write the decay equation for tritium:

$$^3_1\text{H} \; \longrightarrow$$ [2]

(iii) Use the information above to explain why people in the vicinity of the tubes are in no danger from the β$^-$ particles. [2]

....

....

....

Q9 Two technicians measured the count rate, C (in counts per second), over 1000 s, from a freshly produced radioactive sample. They plotted a graph of $\ln C$ against time.

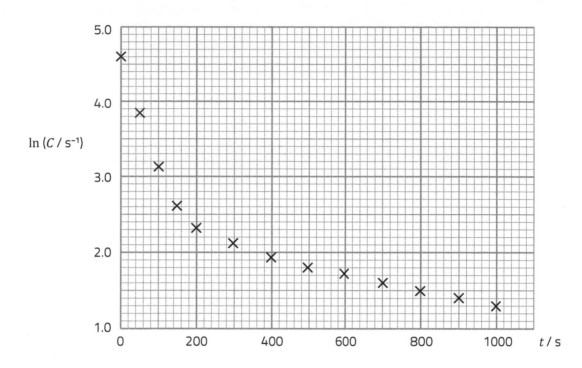

Dominic said that there must be two different radioisotopes in the sample and that after about 400 s, the one with the shorter half-life had decayed to a negligible amount.

(a) State how the results support Dominic's suggestion. [2]

....

....

....

(b) Use the results after 400 s to determine, for **isotope 2** (the one with the longer half-life):

(i) the decay constant. [2]

....

....

....

(ii) the count rate received at time $t = 0$. [2]

....

....

....

(c) Determine the initial count rate from **isotope 1** (the one with the shorter half-life). [2]

....

....

....

Q10 The radioactive nuclide, protoactinium-234, $^{234}_{91}Pa$, is sometimes used in schools for half-life determination because of its very short half-life. $^{234}_{91}Pa$ is produced by the decay of an isotope of thorium (Th), which is formed by the alpha decay of an isotope of uranium (U), which has a proton number of 92.

A website gives the half-life of ^{234}Pa as 1.17 minutes.

(a) Complete the decay equations to show the formation of protoactinium-234 from uranium: [3]

$$U \longrightarrow Th \quad +$$

$$Th \longrightarrow {}^{234}_{91}Pa \quad +$$

(b) In a classroom experiment, a student measured an initial count of 470 ± 22 over a period of 10 seconds. In a second reading 3.0 minutes later, the count was 86 ± 9. A preliminary measurement showed that background radiation was negligible compared to these readings.

Evaluate whether the student's results are consistent with the website data. [5]

....

....

....

....

....

....

....

(c) Describe how the student could obtain better data to compare with the website value of half-life. Briefly describe the method of analysis.

[Note that, with this apparatus, it is not possible to obtain an initial count rate greater than about 500 in 10 seconds.] [3]

....

....

....

....

....

Q11 A beam of β particles is passed through two narrow slits and is detected using a GM tube.

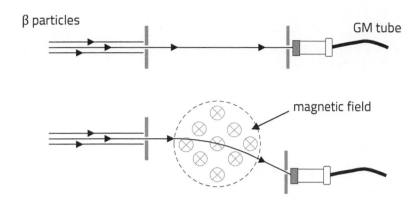

When a magnetic field is applied between the slits, the second slit has to be moved downwards in order to detect the β particles.

(a) Explain this observation and deduce whether the β particles are β⁻ or β⁺. [4]

....

....

....

....

....

(b) Explain how the observations would be different if the beam consisted of α particles or γ photons. [3]

....

....

....

....

....

Question and mock answer analysis

Q&A 1

A group of students investigated two radioactive sources. They first measured the background count over a period of 1 minute using a radiation detector.

Result: Background count in 1 minute = 24.

(a) They set up the radiation detector 10 cm away from a beta (β) source and measured the count over 1 minute. The area of radiation detector was 1.0 cm².

Result: Count in 1 minute = 864.

Estimate the activity of the beta source, explaining your reasoning. [4]

(b) The students then investigated the absorption in lead of the radiation from a gamma source. They inserted a series of 0.50 cm thick pieces of lead between the source and the detector and recorded the corrected count, C, in 1 minute. Their results were as follows:

Absorber thickness, x / cm	C
0.00	572
0.50	308
1.00	233
1.50	144
2.00	81
2.50	67
3.00	40

The expected relationship is $C = C_0 e^{-\mu x}$ where C_0 is the count due to source with no absorber, and μ is a constant.

(i) Show that a plot of $\ln C$ against x is expected to be a straight line. [2]

(ii) Use the grid to plot a graph of $\ln C$ against x and draw a suitable line. Error bars are not required. [8 cm × 12 cm grid supplied] [4]

(iii) Determine a value for μ and give an appropriate unit. [3]

(iv) The source and detector separation was a constant 10.0 cm. Explain, using an example, why the separation needed to be a controlled variable. [2]

What is being asked

Part (a) is a synoptic question from Component 2 requiring application of the inverse square law of radiation. It also requires the handling of background corrections. This is an AO2 question. Part (b) is an experimental analysis based on a specified practical. The theory does not form part of the content of this topic; hence the expected relationship is given. Parts (i) and (ii) are standard AO2. Part (iii) is classed as AO3 because the method of using the results to obtain the answer is not flagged up. Part (iv) relates to the design of the experiment and is AO3.

Mark scheme

Question part			Description	AOs			Total	Skills	
				1	2	3		M	P
(a)			Subtraction of background [1] Conversion of time units [1] Application of $4\pi r^2$ [1] Activity = 17 kBq [1] Allow ecf on 864, using minutes and πr^2		4		4	1 1	4

(b)	(i)	Taking logs correctly, e.g. $\ln C = \ln C_0 - \mu x$ [1] Clear comparison with $y = mx + c$ [1]		2		2	2	
	(ii)	Linear axes labelled with unit on x axis and no unit on $\ln C$ axis [1] Scales chosen so points occupy at least 50% of each axis [1] Points plotted correctly within <1 square [1] Best fit straight line (by eye) drawn [1]		4		4		4
	(iii)	$\dfrac{\Delta y}{\Delta x}$ used for gradient (ignore sign) [1] Widely separated points on graph used. [1] $\mu = 0.86$ cm^{-1} **unit** [tolerance of ± 0.05] [1]		3	3	3	1	3
	(iv)	Inverse square law referred to or implied (accept: the count rate is lower at greater distances <u>because</u> radiation spread out) [1] Calculation, e.g. at 20 cm expect 143 counts with no absorber [1]		2		2	1	2
Total			0	10	5	15	6	13

Rhodri's answers

(a) Corrected count = 842 ✓

Fraction = $\dfrac{1.0 \text{ cm}^2}{\pi \times 10^2}$ = 3.18 × 10⁻³ ✗

so 3.18 × 10⁻³ A = 842 ✗

so A = 260 000 Bq ✓ ecf

MARKER NOTE
Rhodri corrected for background (first mark) but not for time and he incorrectly calculated the area of a 10 cm sphere. However, he obtained the last mark ecf.

2 marks

(b)(i) $\log C = \log C_0 - \mu x$ ✓

so a straight line ✗ (not enough)

MARKER NOTE
A clear comparison with the equation of a straight line needed for the second mark.

1 mark

(ii)
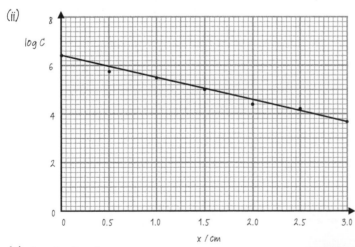

✓
✗
✓
✗

MARKER NOTE
Linear scales and correct labelling gives the first mark. Log is accepted for ln or log$_e$. The second mark is withheld because the points occupy less than half the vertical axis.
Plots are correct – third mark. The line mark is not given because there are more points below the line than above – Rhodri has just joined the first and last points.

2 marks

(iii) $\log C = \log C_0 - \mu x$.

$\log C_0 = 6.35$; when $x = 2$, $\log C = 4.39$ ✓

So 4.60 = 6.35 − 2μ

$\mu = \dfrac{6.35 - 4.39}{2}$ ✓ = 0.88 (2 dp) ✗ (unit)

MARKER NOTE
The use of the equation is equivalent to using the gradient, as long as points from the line are used. The points are widely spaced. The unit is missing so the last mark is not given.

2 marks

(iv) The radiation spreads out so the further away, the lower the counts ✓, so it wouldn't be a fair test. e.g. if it is moved to 15 cm the count would be lower. ✗ (not enough)

MARKER NOTE
The first mark is clearly given (radiation spreading out). A calculation is needed for the second mark.

1 mark

Total	8 marks / 15

Ffion's answers

(a) At 10 cm the radiation spreads out to
area = $4\pi \times 10^2$ = 1257 cm^2 ✓
So activity = 1257 × 864 ✗
= 1.09×10^6 counts per min
= 18 kBq ✓✓ ecf

MARKER NOTE
The only mark that Ffion missed is the correction for background (subtracting 24). In fact the examiner would have given her the marks for the previous line because she gave a correct unit for activity there.

3 marks

(b)(i) Equation of a straight line is y = μx + c.
If $C = C_0 e^{-\mu x}$, $\ln C = \ln C_0 - \mu x$ ✓
This has the same form as y = mx + c
if we plot ln C on the y axis.
The gradient is −μ ✓

MARKER NOTE
Ffion hits both marking points here and gains full marks.

2 marks

(ii)

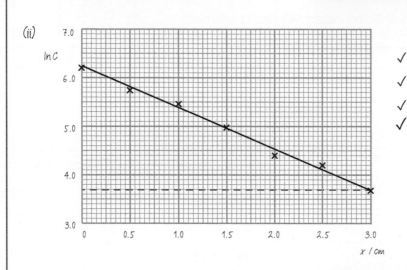

✓
✓
✓
✓

MARKER NOTE
Full marks: Ffion has chosen a ln C scale so that the points occupy more than half the axis.
Her line has a gradient which reflects the points; the points are scattered equally above and below the line.
The dotted line is for part (iii).

4 marks

(iii) μ = gradient = $\dfrac{6.25 - 3.69}{3.0 \text{ cm}}$ ✓✓(bod)
= 0.85 cm^{-1} ✓

MARKER NOTE
Ffion has made two mistakes of sign in what she writes. As she says in (b)(i), the gradient is −μ and her expression is minus the gradient. However, these are not penalised. Her unit is correct.

3 marks

(iv) If the distance increases, the count rate will decrease
because of the inverse square law ✓
e.g. if it is doubled, the count will become $\frac{1}{4}$.
So to make a comparison the distance needs
to be made the same. ✗ (not enough)

MARKER NOTE
The second mark is a difficult one to achieve. An answer such as, "20 cm would give a count of only 77 counts with 0.5 cm of lead – but not because of absorption," would get the mark.

1 mark

Total   **13 marks /15**

Section 7: Particles and nuclear structure

Topic summary

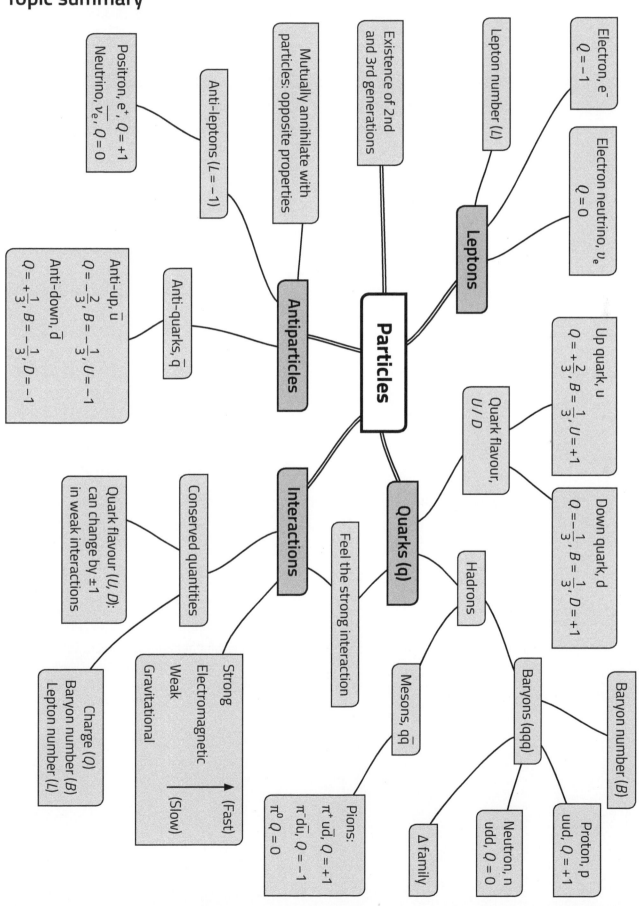

Q1 Electrons can be described as *fundamental particles*. Protons are *composite particles*. Explain this difference for these particles. [2]

....

....

Q2 Particle X can interact through the strong interaction. What can you conclude about X? [1]

....

Q3 The following is a selection of first-generation particles and antiparticles:

electron (e^-)   proton (p)   anti-neutrino ($\overline{v_e}$)   pi+ meson (π^+)   anti-neutron ($\overline{n}$)   positron (e^+)

Identify which of these particles can interact via each of the strong, electromagnetic and weak forces. [2]

Strong: ....

Electromagnetic: ....

Weak: ....

Q4 Neutral pions (π^0) consist of a mixture of $u\overline{u}$ and $d\overline{d}$. They decay into two photons with a lifetime of about 10^{-16} s.

(a) State which interaction is responsible for this decay. [1]

....

(b) Give two reasons for your answer. [2]

(i) ....

(ii) ....

Q5 (a) Name and write the symbol for the anti-particle of each of the following particles: [3]

(i) e^- ....

(ii) n ....

(iii) v_e ....

(b) Eurig says that π^- is the anti-particle of π^+. Determine whether Eurig is correct. [2]

....

....

....

....

Q6 The Δ^{++} is a first-generation particle.

(a) Keira says that Δ^{++} must be a baryon. Explain why she is correct and give its quark structure. [2]

....

....

....

(b) The Δ^{++} has a lifetime of about 10^{-24} s. It decays into a proton with the emission of another first-generation particle.

(i) State the interaction which is responsible for the decay and give a reason for your answer. [2]

....

....

(ii) Momentum and energy are conserved in this decay. State the other quantities which are conserved in this sub-atomic particle decay. [2]

....

....

(iii) Write the equation for this decay and show how each of the quantities you named in (ii) is conserved. [3]

....

....

....

....

....

Q7 Sub-atomic particles can be classified as *leptons*, *baryons* and *mesons*. The positive pion, π^+, decays into a positron, e^+, and a neutrino, ν_e.

(a) Classify the particles, π^+, e^+ and ν_e. [2]

....

....

....

....

(b) Explain which properties of particles are conserved in this decay, in addition to momentum and energy. [3]

....

....

....

....

(c) State a particle property which is not conserved in this decay and explain your answer briefly. [1]

....

....

Q8 Two protons collide at high energy. The following reaction cannot occur.

$$p + p \rightarrow \Delta^+ + e^- + \pi^+$$

where Δ^+ is a first-generation baryon.

(a) Explain which of the conservation laws this reaction would violate. [3]

....

....

....

....

....

(b) Usually a Δ^+ decays into a nucleon and a pion, e.g. $\Delta^+ \rightarrow p + \pi^0$, with a lifetime of $\sim 10^{-24}$ s. Rarely, however, the following decay is observed:

$$\Delta^+ \rightarrow p + \gamma$$

Explain why this is a much slower decay. [2]

....

....

....

Q9 A neutron and a positive pion collide at high energy. A student suggests that the following reaction might be observed:

$$n + \pi^+ \rightarrow \Delta^{++} + e^-,$$

where the Δ^{++} is a first-generation baryon.

Explain whether this reaction would violate the conservation of lepton number, charge and baryon number. [3]

....

....

....

....

Q10 Quarks are said to have a *baryon number* of $\frac{1}{3}$. Explain, using examples, how this is consistent with each of the following:

(a) Baryons each consist of three quarks. [2]

....

....

....

(b) Mesons each have a baryon number of 0 (zero). [2]

....

....

....

Q11 The Sun emits the Solar Wind, which is a stream of particles, mainly electrons and protons. The nuclear reactions in its core also produce a stream of neutrinos. If these particles were to hit a piece of lead, whilst the electrons and protons would have a range of less than 1 cm, the range of the neutrinos has been estimated as 1 light year.

Explain this difference by considering the kinds of interaction in which the three particles can engage.

[4]

Q12 The laws of conservation of energy and momentum apply to all collisions. Write a brief account of the **additional** conservation laws which apply to interactions between sub-atomic particles.   [6 QER]

Q13 When an isolated particle decays, the conservation of mass and energy requires that the total mass of the resulting particles is lower than the mass of the decaying particle.

The following table contains the masses of **all** first-generation particles as multiples of the electronic mass, m_e.

Particle	Symbol(s)	Type	Mass / m_e
neutrino	ν	lepton	$<10^{-5}$
electron	e^-	lepton	1
pion	$\pi^+ / \pi^0 / \pi^-$	meson	207
proton	p	baryon	1836
neutron	n	baryon	1839
delta	$\Delta^{++} / \Delta^+ / \Delta^0 / \Delta^-$	baryon	2411
rho	$\rho^+ / \rho^0 / \rho^-$	meson	1516

Second- and third-generation particles of each type have greater masses.

Use this information to explain each of the following statements. For some of the statements you will need to consider other conservation laws.

(a) The Δ^+ particle can decay into a proton and a π^0 but not a proton and a ρ^0. [2]

....

....

....

(b) The π^0 can decay into an electron, a positron and a photon. [2]

....

....

....

(c) The neutron can only decay by means of the weak interaction. [3]

....

....

....

(d) The proton is a stable particle, i.e. it does not decay. [2]

....

....

....

....

Question and mock answer analysis

Q&A 1   Particle Y decays, producing a neutral pion, π^0, a positron, e^+, and a lepton, Z, with no charge.

(a)   Imran says that particle Y could be a positive baryon. Josephine thinks that it could be a neutral meson. Evaluate which, if either, is correct.   [4]

(b)   Jemima says that Z must be a neutrino. Evaluate this statement.   [2]

(c)   Explain which interaction, strong, weak or electromagnetic, is responsible for the decay of particle Y.   [2]

(d)   Explain how a measurement of the mean lifetime of Y particles would provide supporting evidence for your answer to (c).   [2]

What is being asked

This is quite a tricky question. Parts (a) and (b) require candidates to use conservation laws in particle physics to come to conclusions about the nature of particles. Because the question does not mention conservation laws, these two parts of the question are classed as AO3. Parts (c) and (d) test the candidates' knowledge of the different kinds of interactions that sub-atomic particles undergo. The basis of the analysis is given in part (c) so this is an AO2 question; in part (d) the candidates are pointed very much in the direction of recalled properties of the weak interaction and so it is AO1.

Mark scheme

Question part			Description	AOs			Total	Skills	
				1	2	3		M	P
(a)			Statement of any conservation law (charge, baryon number or lepton number) [1] e.g. 'charge is conserved'			4	4		
			Correct application to the decay [1]						
			Correct application of a second law [1]						
			Identification of particle, following correct reasoning, and evaluation [1]						
(b)			The only leptons with no charge are neutrinos [or anti-neutrinos] [1]		2		2		
			Z cannot be an antineutrino because total lepton number would be −2 (which Y cannot have) so Jemima is correct [1]						
(c)			The reaction involves a neutrino [1]		2		2		
			[Neutrinos only interact via the weak interaction] **hence** weak [1]						
(d)			Weak decays take longer than strong or e-m [1]	2			2		
			If decay time longer than (e.g.) 10^{-10} s then weak [1]						
Total				2	2	6	10		

Rhodri's answers

(a) Y must have a positive charge because charge is always conserved ✓ and the product particles have charges of 0 + 1 + 0 = 1. ✓ So Josephine cannot be correct but Imran could be right.

There is a meson on the right so there must be a meson on the left ✗ so the particle is a positive meson (not clear ✗)

MARKER NOTE
Rhodri handles the conservation of charge well and correctly indicates that Y must have a positive charge. He thus obtains the first two marks on the scheme. He mistakenly thinks that mesons are conserved; so that, although his conclusion is correct it is arrived at by faulty reasoning.

2 marks

(b) A neutral lepton must be a neutrino ✓
so Jemima is correct (not enough ✗)

MARKER NOTE
The marking scheme allows an easily accessible first mark which Rhodri achieves. However, he does not rule out an anti-neutrino and so was not given the second mark.

1 mark

(c) The reaction involves a neutrino ✓, so it must be weak ✓ (just)

MARKER NOTE
The marking scheme generously allowed Rhodri's answer – the section in brackets in the scheme was not required.

2 marks

(d) Weak interactions are less likely to happen than strong. They have shorter range than electromagnetic interactions – so this reaction is unlikely to happen. ✗

MARKER NOTE
Rhodri's analysis relates to a collision interaction rather than to a decay. Hence he talks about the probability of a reaction occurring. This doesn't attract either point of the mark scheme.

0 marks

| Total | **5 marks /10** |

Ffion's answers

(a) Y cannot be a lepton because there are two leptons on the right. ✓

Y cannot be a baryon because there are no baryons on the right. ✓

Y must have a positive charge because the total charge on the right is +1. ✓

So the particle is a positive meson. (not clear ✗)

MARKER NOTE
Ffion doesn't actually state any conservation law explicitly but applies all three correctly. So the examiner has awarded the first mark slightly generously by implication. On the other hand, Ffion has not actually answered the question fully; to be awarded the last mark she needed to say to what extent each person was correct..

3 marks

(b) The lepton number of e^+ is –1, so the lepton number of Z is +1 so it must be either an electron or a neutrino ✓

Z is neutral so it must be a neutrino. ✓

MARKER NOTE
Ffion sets out her answer differently from the mark scheme but clearly addresses both marks correctly.

2 marks

(c) Leptons don't feel the strong force (because they have no quarks). Neutrinos are neutral so they don't feel the e-m force so it must be weak. ✓✓

MARKER NOTE
This answer by Ffion is much better than Rhodri's but she cannot get more than 2 marks. Again, her setting out was very different from that in the mark scheme.

2 marks

(d) Strong decays happen very quickly (about 10^{-24} s). Electromagnetic decays take longer (about 10^{-16} s). Weak decays take the longest ✓ ✗

MARKER NOTE
A very good start to the answer. Unfortunately she only achieves the first marking point – with her last statement. Unaccountably she misses out an estimate of the time for a weak decay.

1 mark

| Total | **8 marks /10** |

Q&A 2 Particles which contain quarks or anti-quarks are called hadrons. Sub-families of the hadrons are *baryons* and *mesons*.

(a) Compare baryons and mesons in terms of quark make-up and a conservation law. [3]

(b) Two high-energy protons collide and undergo the following reaction:

$$p + p \rightarrow p + X + \pi^+$$

where X is a first-generation particle.
Use conservation laws to identify particle X. Give your reasoning. [4]

Component 3 Practice questions

What is being asked

This is intended to be a straightforward question. Part (a) requires you to remember details of the nature and properties of baryons and mesons. This is bookwork and hence AO1. Part (b) asks you to apply your knowledge to a given reaction. The mark scheme allows this to be done in terms of baryons or quarks. Notice that a simple identification of X will not score, even if correct, because the question asks for your reasoning. The basis of the analysis is given, so this is an AO2 question.

Mark scheme

Question part			Description	AOs			Total	Skills	
				1	2	3		M	P
(a)			Baryons are composed of 3 quarks [1]	3			3		
			Mesons are composed of a quark and an anti-quark [1]						
			[In any interaction] the baryon number [or number of baryons] is conserved but the number of mesons is not [1] **Both needed**						
(b)			Protons are baryons and π^+ is a meson [1]		4		4		
			Baryon number = 2, so X is a baryon [1]						
			Charge is conserved so X is neutral / uncharged [1]						
			Uncharged [first-generation] baryon is neutron (accept Δ^0) [1]						
			Alternative (in terms of quarks)						
			proton composition = uud; $\pi^+ = u\bar{d}$ (✓)						
			To conserve charge, d must be created alongside $\bar{d}$ (✓) (Accept: quark flavour is conserved)						
			Quarks on both sides = 4u + 2d (✓)						
			∴ Composition of X = udd hence neutron (✓)						
Total				3	4		7		

Rhodri's answers

(a) Baryons are uud or udd (where u is an up-quark and d is a down-quark). ✗

Mesons only have two quarks – one of which is an anti-quark, e.g. u$\overline{\text{d}}$ ✓ [bod]

Baryon number is conserved but mesons have no conservation laws. ✓ [bod]

MARKER NOTE
Rhodri answers in terms of specific baryons without naming them, rather than a general answer, which is not credited. His meson answer is also not ideal, not making it clear that it is a quark / anti-quark pair, but the examiner gives him the benefit of the doubt. He clearly states that baryons are conserved; mesons do obey the law of conservation of charge but his answer is close enough for the examiner to award the last mark.

2 marks

(b) Protons are uud quarks

To conserve quarks X + π$^+$ must be uud. ✓

π$^+$ is u$\overline{\text{d}}$ ✓, so the quark composition of X is udd, ✓ which is neutral because u = $+\frac{2}{3}$ and d = $-\frac{1}{3}$. ✗ not enough

MARKER NOTE
Rhodri goes down the quark route and does rather well. The first mark he obtains is actually the second one on the mark scheme. The next mark is for both the proton and pion structures. He applies the conservation of quark flavour, although this is not clearly expressed for the third mark. He just misses out on the last mark because he doesn't actually say that X is a neutron.

3 marks

Total	5 marks /7

Ffion's answers

(a) Baryons have 3 quarks, e.g. proton = uud ✓

Mesons have a quark and an antiquark, e.g. π$^+$ = u$\overline{\text{d}}$ ✓

In collisions or decays, the baryon number is conserved ✗ [not enough]

MARKER NOTE
Ffion gives a textbook answer for the structure of both baryons and mesons. Her examples, the proton and positive pion, are not required for the first two marks.
She correctly mentions the conservation of baryon number. To nail the final mark she needed to say that there is no law of conservation of mesons.

2 marks

(b) In this reaction the protons are baryons, so the number of baryons must stay at 2. So X must be a baryon. ✗ [not enough] ✓

The meson has taken the positive charge ✓ [bod] so X must be neutral.

So X is a neutron. ✓

MARKER NOTE
Ffion clearly identifies protons as baryons. In order to obtain both the first two marks, she should have said that the π$^+$ was a pion, so is not counted in the baryon number, leaving X to be a baryon.
She applies the conservation of charge; this could have been done more clearly, hence the bod comment. She correctly finishes off with identifying the neutral baryon as a neutron.

3 marks

Total	5 marks /7

Section 8: Nuclear energy

Topic summary

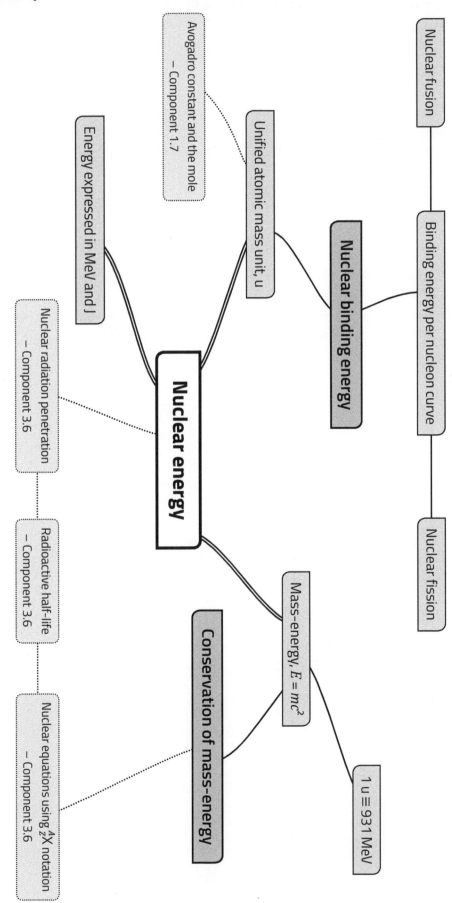

Nuclear energy

Nuclear binding energy

Conservation of mass-energy

Nuclear fusion

Binding energy per nucleon curve

Nuclear fission

Avogadro constant and the mole
— Component 1.7

Unified atomic mass unit, u

Energy expressed in MeV and J

Nuclear radiation penetration
— Component 3.6

Radioactive half-life
— Component 3.6

Nuclear equations using $_Z^A X$ notation
— Component 3.6

Mass-energy, $E = mc^2$

$1\,u \equiv 931\,\text{MeV}$

Q1 When asked to define the *binding energy* of a nucleus, Julia wrote:

This is the energy needed to hold the particles of a nucleus together.

Octavia said that this could not be right because that would mean that a nucleus would have a greater mass than the protons and neutrons in it.

(a) Explain Octavia's comment. [2]

....

....

....

(b) Write a correct definition of nuclear binding energy. [2]

....

....

....

Q2 A website gives the mass of a neutral $_1^1\text{H}$ atom as 1.007 825 032 u and the ionisation energy (the energy needed to remove the electron) as 13.6 eV.

Use the following data to evaluate whether the sum of the masses of a proton and an electron is different from the atomic mass to this number of significant figures:

$u = 1.66 \times 10^{-27}$ kg $e = 1.60 \times 10^{-19}$ C $c = 3.00 \times 10^8$ m s^{-1} [3]

....

....

....

....

....

Q3 A data book gives the following mass data in u:

electron: 0.000 549   proton: 1.007 276   neutron: 1.008 665 $_2^4\text{He}$ atom: 4.002 604

Use this information to calculate:

(a) the binding energy of a $_2^4\text{He}$ atom [3]

....

....

....

....

(b) the binding energy per nucleon of a $_2^4\text{He}$ atom [1]

....

Q4 The intensity of electromagnetic radiation from the Sun incident on the Earth's atmosphere is 1370 W m^{-2}. The radius of the Earth's orbit is 1.50×10^{11} m.

(a) Calculate the power emitted by the Sun as electromagnetic radiation. [2]

....

....

....

(b) It is commonly stated that the Sun loses 4 million tonnes of mass each second due to the output of radiation. Evaluate this statement. [1 tonne = 1000 kg] [2]

....

....

....

Q5 A science data book gives the following data for uranium-235 which decays by α emission:

Atomic mass = 235.043 930 u; density = 18.8×10^3 kg m^{-3}; half-life = 7.1×10^8 years.

(a) Calculate the activity of 1.0 kg of pure $^{235}_{92}$U. [4]

....

....

....

....

....

....

(b) Michael and Jonathan were discussing how a 1.0 kg lump of $^{235}_{92}$U would feel. They agreed that it might be warm to the touch. Michael suggested you wouldn't notice this.

(i) Explain why it might be warm. [2]

....

....

....

....

(ii) Discuss Michael's suggestion. [4]

Additional data: $m(^{231}_{90}\text{Th}) = 231.036\ 304$ u; $m(^4_2\text{He}) = 4.002\ 604$ u

....

....

....

....

....

....

Q6 Tritium, ^3_1H, is a radioactive isotope of hydrogen. It is hoped to use tritium in a nuclear fusion reactor.

Tritium is made by bombarding the lithium isotope ^6_3Li with neutrons in a specially designed fission reactor. The reaction also produces one other nuclide in the reaction.

The grid shows the binding energy per nucleon of various nuclides given to the nearest 0.1 MeV nuc^{-1}.

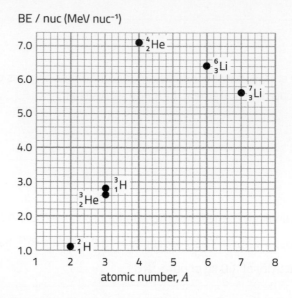

(a) Complete the equation of the nuclear reaction between ^6_3Li and a neutron which produces tritium:

$$^6_3\text{Li} + ^1_0\text{n} \longrightarrow$$

[2]

(b) Tritium decays by α, β^- or β^+ emission into one of the other nuclides identified on the chart.

Write the decay equation for tritium:

$$^3_1\text{H} \longrightarrow$$

[2]

(c) In the proposed reaction in the fusion reactor, tritium reacts with deuterium, another isotope of hydrogen, ^2_1H, to produce ^4_2He and one other particle.

Determine the energy released in the fusion of 1.0 kg of an appropriate mixture of tritium and deuterium.

[5]

Q7 The Sun, a middle-aged star, derives its energy from the fusion of hydrogen to helium in a process which can be summarised as:

$$4\,^1_1\text{H} \longrightarrow\ ^4_2\text{He} + 2\,^0_{-1}\text{e}$$

where ^1_1H and ^4_2He are the atomic symbols. This stage in the life-cycle of the Sun is expected to last 9×10^9 years in total.

The Sun will later go through a stage of 'helium-burning' in which the helium produced by hydrogen fusion reacts to form carbon in the so-called triple-alpha process:

$$^4_2\text{He} +\ ^4_2\text{He} +\ ^4_2\text{He} \longrightarrow\ ^{12}_6\text{C}$$

Data: $m\,(^1_1\text{H}) = 1.007\ 825$ u; $m(^4_2\text{He}) = 4.002\ 604$ u; $m\,(^{12}_6\text{C}) = 12$ u (exactly); $m\,(^0_{-1}\text{e}) = 0.000\ 549$ u

(a) Show that the fusion of four ^1_1H atoms to ^4_2He yields 25.7 MeV. [2]

(b) Calculate the energy released in the triple-alpha process. [2]

(c) During the 'helium-burning' phase the Sun will have 10× its current diameter; its surface temperature will be 90% its current kelvin value.

(i) Show that the luminosity of the Sun will be approximately 65× its current value. [2]

(ii) Use the information at the start of the question and your answers to estimate the length of time that the helium-burning phase is able to last. Show your reasoning. [4]

Q8 Elements with nucleon numbers greater than 56 are produced in supernova explosions and in the merger of neutron stars. In these very dense conditions with large numbers of free neutrons, nuclei such as iron-56 ($^{56}_{26}$Fe) absorb neutrons and then undergo β^- decay in a process which builds up heavier nuclei.

The production of $^{235}_{92}$U from $^{56}_{26}$Fe can be summarised by:

$$^{56}_{26}\text{Fe} + \text{.............} \, ^{1}_{0}\text{n} \longrightarrow \, ^{235}_{92}\text{U} + \text{.............} \, ^{0}_{-1}\text{e} + \text{.............} \, ^{0}_{0}\overline{\nu}_e$$

Complete the equation with the correct numbers of particles and explain your reasoning. [4]

....

....

....

....

....

Q9 The mass of a $^{4}_{2}$He atom is 4.002 604 u. The mass of a $^{8}_{4}$Be atom is 8.005 305 u. $^{8}_{4}$Be decays (with a very short half-life of 10^{-16} s) into two $^{4}_{2}$He atoms.

(a) Explain why $^{8}_{4}$Be is able to decay to $^{4}_{2}$He in this way and suggest why the decay has such a short half-life. [2]

....

....

....

(b) If the $^{8}_{4}$Be is stationary, the two resulting $^{4}_{2}$He nuclei are seen to move off in opposite directions.

(i) Explain this observation. [2]

....

....

....

(ii) Calculate the speeds of the $^{4}_{2}$He nuclei. [The mass of electrons can be ignored.] [4]

....

....

....

....

....

....

Question and mock answer analysis

Q&A 1

In a prototype nuclear fusion reactor, a $^3_1$H nucleus and a $^2_1$H nucleus collide head-on at high speed. They combine to produce a stationary unstable nucleus A, which decays with a half-life of 7.6×10^{-22} s into a nucleus of $^4_2$He and another particle B. This is illustrated in the following diagram:

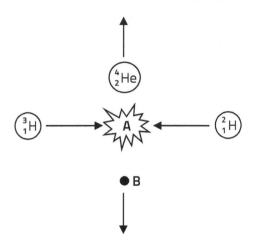

Component 3 Practice questions

(a) Identify the unstable nucleus A and particle B, briefly explaining your answer. [1]

(b) Identify the nuclear interaction responsible for the decay of A **giving two reasons** for your answer. [2]

(c) Explain why the two hydrogen nuclei need to approach at high speed for these reactions to occur. [2]

(d) Alex and Eirian discussed these reactions in terms of mass/energy. Alex claimed the total mass of the $^3_1$H and $^2_1$H must be more than the mass of A. Eirian disagreed and said that by conservation of mass-energy, these masses must be equal.
Discuss who, if either, is correct. [2]

(e) In one such reaction, the intermediate nucleus A is stationary. The masses of some of the particles are given:

$m(A) = 5.010\ 959$ u; $m(^4_2\text{He}) = 4.001\ 505$; $m(B) = 1.008\ 665$ u

   (i) Calculate the energy released in the decay of A. Give your answer in J. [3]

   (ii) Explain why the speed of particle B is 4.0 times that of the $^4_2$He nucleus. [2]

   (iii) Use (i) and (ii) to determine the speed of the $^4_2$He nucleus. [3]

   (iv) In order for the fusion of the hydrogen nuclei to occur, they have to approach within 1.0×10^{-14} m. By considering energy conservation, calculate the total kinetic energy of these two particles before the collision and hence estimate the necessary temperature of the gas in the reaction chamber. [5]

What is being asked?

The question involves two reactions, nuclear fusion and an unusual radioactive decay. It brings many areas of the A Level Physics specification into play. Section 3.8 relies heavily on the knowledge of other sections of the specification, so many questions are synoptic in nature. Part (a) relies on the concepts of Section 3.6. Particle physics (3.7) is used in parts (b) and (c). Part (d) is an AO3 question which is purely Section 3.8. Part (e) brings in momentum conservation in (ii), kinetic energy in (iii). The last part, (e)(iv), another AO3 question, is another synoptic part involving electric fields (2.5) and kinetic theory (1.7) from workbook 1.

Mark scheme

Question part			Description	AOs			Total	Skills	
				1	2	3		M	P
(a)			A = $^5_2$He (nucleus), **and** B = neutron Correct reasoning including proton and neutron numbers in $^5_2$He (or equiv) [1]		1		1		
(b)			Strong interaction / force because: • Half-life {so short / typical of strong} [1] • Only hadrons/quarks involved **or** no change in quark flavour [1]			2	2		
(c)			Nuclei [both] positively charged so repel [1] Need to approach closely for fusion to occur [1]	2			2		
(d)			The initial <u>kinetic energy</u> (of the $^3_1$H and $^2_1$H) also has mass. [1] (Mass/energy is conserved) so the mass (of A) is more than the total (rest) mass of $^3_1$H and $^2_1$H; hence neither is correct. [1]			2	2		
(e)	(i)		Mass loss = 7.89×10^{-4} (u) [1] [no sign penalty] Energy released = 0.735 MeV [1] = 1.176×10^{-13} J [1]		3		3	3	
	(ii)		The momenta of $^4_2$He and {B/neutron} are equal (and opposite) [1] $m(^4_2$He$) = 4.0 \times m($B$)$ [1]		2		2		
	(iii)		If speed of $^4_2$He = v, KE = $\frac{1}{2}(4.00v^2 + 1.01 \times (4.0v)^2)[\times 1.66 \times 10^{-27}]$ J [1] = $1.67 \times 10^{-26}v^2$ ∴ $1.67 \times 10^{-26}v^2 = 1.176 \times 10^{-13}$ J [1] ∴ Speed of $^4_2$He = 2.65×10^6 m s^{-1} [1]		3		3	2	
	(iv)		Use of $\frac{1}{4\pi\varepsilon_0}\frac{Q_1 Q_2}{r}$ [1] [e.g. 2.30×10^{-14} J seen] Use of conservation of energy [1] Use of $(\frac{3}{2})kT$ with mean particle energy [i.e. 1.15×10^{-14} J ecf] [1] ~800 MK [or 560 MK] obtained [1] ecf Suggestion that a significantly lower temperature is will produce fusion. [1]			5	5	3	
Total				**2**	**9**	**9**	**20**	**8**	

Rhodri's answers

(a) It must be $^5_2$He because it has the right number of protons (2) and neutrons (3). ✗

> **MARKER NOTE**
> B not identified as a neutron.
>
> **0 marks**

(b) It must be a strong interaction because it happens in such a short time – 10^{-22} s. ✓
Second reason ???

> **MARKER NOTE**
> Only one of the two reasons given.
>
> **1 mark**

(c) The strong force has a very short range so the particles have to collide in order for the reaction to happen, which needs a high energy ✓ bod .

> **MARKER NOTE**
> Rhodri has given the information about the distance of approach needed. He hasn't related this to overcoming the repulsion. This might be hinted at with the need for a high energy, but that is not clearly stated.
>
> **1 mark**

(d) The total mass-energy is the same for all three stages of the process as long as we remember to include the kinetic energy of the particles (because there is nowhere else for the energy to go). ✓ So the $^3_1$H and $^2_1$H masses will be more than when they are at rest and Eirian is correct. ✓

> **MARKER NOTE**
> The examiner has used judgement here to award marks for a different interpretation. Rhodri considers the masses of the $^3_1$H and $^2_1$H to include that of the kinetic energy and in that sense, Eirian is correct.
>
> **2 marks**

(e)(i) Mass of products = 5.01017 u
Mass loss = 0.000789 u ✓
= 1.31×10^{-30} kg
∴ Energy released = mc^2
= 1.18×10^{-13} J ✓✓

> **MARKER NOTE**
> Rather than use the conversion factor 931 MeV/u, Rhodri has converted to kg and used $E = mc^2$. This is perfectly acceptable.
>
> **3 marks**

(ii) The momentum of B is equal and opposite to the momentum of the $^4_2$He ✓. B has $\frac{1}{4}$ of the mass so 4× the speed. ✓

> **MARKER NOTE**
> Neatly expressed.
>
> **2 marks**

(iii) Ratio of masses 1:4. Using $\frac{m}{m \times M}$, $^4_2$He gets $\frac{1}{5}$ of the total energy = 2.36×10^{-14} J ✓
So $\frac{1}{2}$ 4.00 × 1.66 × 10^{-27} v^2 = 2.36×10^{-14} ✓
∴ $v = 1.86 \times 10^6$ m s^{-1} ✗

> **MARKER NOTE**
> Rhodri has not used the way in from (ii) but has got off to a good start using ratios and worked out the energy of the ^{4}He nucleus. Unfortunately he has forgotten the ½ in calculating v, so loses the third mark.
>
> **2 marks**

(iv) Energy released = 1.18×10^{-13} J ✗
This is shared between the particles so mean energy = 5.9×10^{-14} J
$E = \frac{3}{2}kT$ ∴ $T = \frac{2 \times 5.9 \times 10^{-14}}{3 \times 1.38 \times 10^{-23}}$ ✓ ecf
= 2.85×10^9 K ✓ecf
Even at lower temperatures (e.g. $\frac{1}{2}$ this) some molecules have enough energy ✓

> **MARKER NOTE**
> Rhodri has not understood this synoptic question which requires knowledge of Component 1. He has used the KE of the daughter particles and hence loses the first two marks. However, his analysis of the relationship between energy and temperature, which are also Component 1 concepts, is spot on.
>
> **3 marks**

> **Total**   **14 marks /20**

Ffion's answers

(a) $^3_1H + ^2_1H \longrightarrow ^5_2He$ (A) $\longrightarrow ^4_2He + ^1_0n$ (B)

This balances so ^5_2He and a neutron ✓

> **MARKER NOTE**
> Not what was expected but Ffion clearly knows what the reactions are, and the explanation implies consistent number of protons and neutrons.
>
> **1 mark**

(b) Only protons and neutrons are involved ✓ (no leptons or photons) and it happens so quickly ✓, so it must be the strong interaction.

> **MARKER NOTE**
> 'Only protons and neutrons' is as good here as 'only baryons'.
>
> **2 marks**

(c) The ^3_1He and ^3_2He have got to get very close for fusion.✓ They are both positively charged so repel each other.✓ If they had low speeds, they wouldn't get close enough.

> **MARKER NOTE**
> Both marking points are clearly made. The last sentence was not required on this occasion.
>
> **2 marks**

(d) Producing A is like a reverse fission reaction. In a fission reaction, there is a mass loss which results in the energy output. ✗ [Not enough]

Hence the mass of the $^3_1He + ^2_1He$ must be less than A and they are both wrong. ✓

> **MARKER NOTE**
> The first marking point is not quite there. The <u>kinetic</u> energy of the hydrogen nuclei needs to be clearly linked to mass-energy.
> The second point is clearly made.
>
> **1 mark**

(e) (i) $\Delta m = 5.010\,959 - 4.001\,505 - 1.008\,665$ ✓

$= 0.000\,789\,u$ ✓

$1\,u = 931\,MeV$

So energy released $= 0.000\,789 \times 931$

$= 0.735\,MeV$ ✓

> **MARKER NOTE**
> Strictly Δm represents the mass gain rather than the mass loss but no penalty is applied on this occasion. Ffion gave the answer in MeV which is accepted.
>
> **3 marks**

(ii) The momentum of B is equal and opposite to the momentum of the ^4_2He ✓. B has $\frac{1}{4}$ of the mass so 4x the speed. ✓

> **MARKER NOTE**
> Neatly expressed.
>
> **2 marks**

(iii) If speed of ^4_2He is v its KE is:

$\frac{1}{2}\,4.00 \times 1.66 \times 10^{-27}v^2 = 3.32 \times 10^{-27}v^2$

and KE of neutron $= 13.28 \times 10^{-27}v^2$

∴ Adding: $1.66 \times 10^{-26}v^2$ ✓ $= 1.18 \times 10^{-13}$ ✓

∴ $\longrightarrow v = 2.67 \times 10^6\,m\,s^{-1}$ ✓

> **MARKER NOTE**
> A very clear answer which ticks all the boxes. Note that Ffion had to do the MeV to J conversion which she didn't do in (e)(i).
>
> **3 marks**

(iv) $PE = 9 \times 10^9 \times \dfrac{(1.6 \times 10^{-19})^2}{1.0 \times 10^{-14}} = 2.30 \times 10^{-14}\,J$ ✓

∴ Initial KE of particles $= 2.30 \times 10^{-14}\,J$ ✓

∴ Using $KE = \frac{3}{2}kT \longrightarrow T = 1.11 \times 10^9\,K$ ✓ecf

> **MARKER NOTE**
> Ffion used the potential energy formula correctly and (by implication) energy conservation for the first two marks. Her calculation of temperature was not correct because she didn't use the mean kinetic energy, so she lost the third mark. The difficult last mark required a comment taking into account energy distribution.
>
> **3 marks**

> **Total** **17 marks /20**

Section 9: Magnetic fields

Topic summary

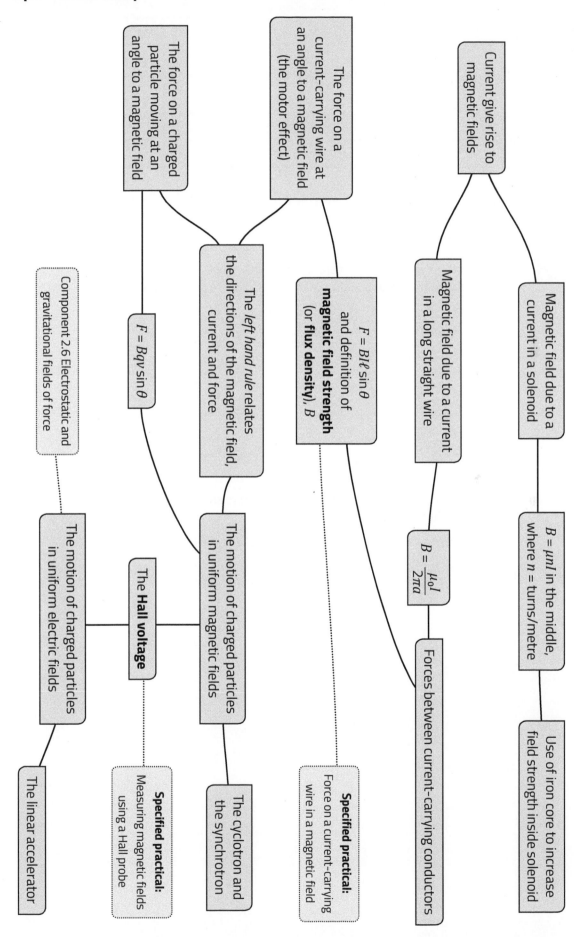

Current give rise to magnetic fields

The force on a current-carrying wire at an angle to a magnetic field (the motor effect)

The force on a charged particle moving at an angle to a magnetic field

The *left hand rule* relates the directions of the magnetic field, current and force

$F = BI\ell \sin\theta$ and definition of **magnetic field strength (or flux density)**, B

$F = Bqv\sin\theta$

Component 2.6 Electrostatic and gravitational fields of force

Magnetic field due to a current in a long straight wire

$B = \dfrac{\mu_0 I}{2\pi a}$

Magnetic field due to a current in a solenoid

$B = \mu nI$ in the middle, where n = turns/metre

Use of iron core to increase field strength inside solenoid

Forces between current-carrying conductors

Specified practical: Force on a current-carrying wire in a magnetic field

The motion of charged particles in uniform magnetic fields

The **Hall voltage**

The motion of charged particles in uniform electric fields

The cyclotron and the synchrotron

Specified practical: Measuring magnetic fields using a Hall probe

The linear accelerator

Q1 When placed in a magnetic field, B, a current-carrying wire experiences a force, F, given by:

$$F = BI\ell\sin\theta$$

(a) Identify the quantities I, ℓ and θ. [1]

....

....

(b) Use the equation to express the tesla (T) in terms of the base SI units kg, m, s and A. [2]

....

....

....

(c) State the name of the rule that links the direction of F with those of the current and field. [1]

....

Q2 A clockwise current of 2.5 A is maintained in a triangular loop of wire by a battery (not shown). A uniform magnetic field of 0.030 T, at right angles to the plane of the loop, is applied as shown:

(a) Add arrows to the centres of each side of the loop to show the directions of the forces on the sides due to the uniform magnetic field. [2]

(b) Calculate the magnitudes of the forces on the following sides: [5]

(i) AB ....

....

(ii) BC ....

....

(iii) CA ....

....

(c) Show that the resultant force on the loop is zero. [3]

....

....

....

....

....

Q3 A uniform magnetic field of magnitude B is applied to the right as shown, so that it surrounds the triangular loop shown, which carries current I.

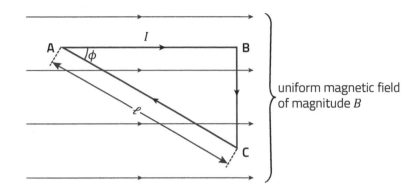

(a) Ella suggests that the resultant force on the loop is zero. Evaluate her suggestion. [4]

....

....

....

....

....

....

....

(b) Even so, Ella argues, the loop will not remain stationary unless it is fixed. Discuss this further claim. [2]

....

....

....

Q4 Three long, parallel wires, P, Q and R are 60 mm apart from each other, as shown in the (cross-sectional) diagram. They each carry a current of 7.5 A into the page.

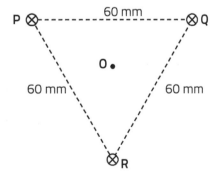

(a) (i) Point O is equidistant from P, Q and R. **Draw arrows labelled P, Q and R on the diagram** to show the directions of the magnetic fields at O due to P, to Q and to R. [2]

(ii) State the magnitude of the resultant magnetic field at point O, justifying your answer briefly. [2]

....

....

....

(b) Stating its direction, determine the force on a 2.0 m length of wire R:

(i) when the current in wire Q is temporarily turned off; [2]

....

....

....

(ii) when the current is restored in Q, so P, Q and R all carry 7.5 A again. [3]

....

....

....

Q5 The diagram shows part of the magnetic field due to a current-carrying solenoid.

(a) The direction of the field in the solenoid is from left to right.

(i) List what else can be deduced about the solenoid's field from the pattern of magnetic field lines. [3]

....

....

....

(ii) Describe how you could investigate the deductions you have made in part (i). [3]

....

....

....

....

(b) State the direction, A or B, of the current, naming the rule that you have used. [1]

....

(c) The solenoid is 60.0 cm long and has 300 turns of wire. Calculate the magnetic flux density, B, at its centre when it carries a current of 4.0 A. [2]

....

....

....

(d) State how you could increase the flux density at the centre, without changing the current, the number of turns or the length of the solenoid. [1]

....

Q6 The diagram shows a beam of electrons passing through a region of uniform magnetic field. The electrons are moving at a speed of 3.0×10^7 m s^{-1} in an arc of a circle of radius 0.040 m.

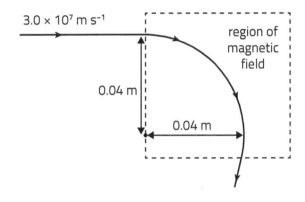

(a) The electrons have been accelerated to this speed from rest. Calculate the accelerating voltage used. [2]

....

....

....

(b) Determine the magnitude and direction of the magnetic flux density. [3]

....

....

....

(c) The electron beam can be made to go straight by applying, as well as the magnetic field, a suitable electric field. Determine its magnitude and direction. [2]

....

....

....

Q7 (a) A particle of mass m and charge q moves in a circular path in a uniform magnetic field, B, perpendicular to the plane of the circle. Starting from the force experienced by the particle, show that the time the particle takes per revolution is:

$$T = \frac{2\pi m}{qB}$$

[3]

....

....

....

....

(b) A simplified diagram of a cyclotron is given below:

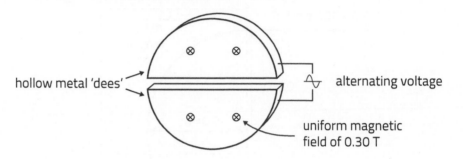

Calculate the frequency of alternating voltage needed if protons are being accelerated in the cyclotron and a magnetic field of 0.30 T is being applied.

[2]

....

....

....

(c) A synchrotron is a particle accelerator whose principle was developed from that of a cyclotron. State two differences between a synchrotron and a cyclotron, and one difference between the ways in which they are used.

[3]

....

....

....

....

....

....

Q8   Here is a simplified diagram of part of a linear accelerator for accelerating protons:

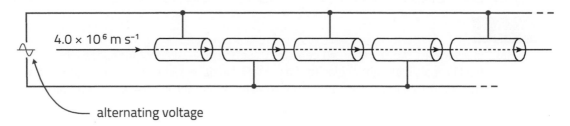

alternating voltage

Protons enter the left-hand tube at a speed of 4.0×10^6 m s^{-1}. As a proton passes from left to right across the gap between two tubes, the accelerating voltage between the tubes is (−)120 kV.

(a)   Calculate a proton's speed after it has entered the right-hand (fifth) tube shown in the diagram. [Mass of proton = 1.67×10^{-27} kg]   [4]

....

....

....

....

....

....

....

(b)   Explain:

(i)   Why an *alternating* voltage is needed.   [2]

....

....

....

(ii)   Why the tubes are of different lengths.   [2]

....

....

....

(c)   State one disadvantage of a linear accelerator over a cyclotron.   [1]

....

Question and mock answer analysis

Q&A 1

(a) In a vacuum, moving charged particles travel in circles at constant speed when a uniform magnetic field is applied at right angles to the particles' initial velocity. In a slice of conducting material carrying a current, the charged particles follow straight paths shortly after a uniform magnetic field is applied, as in the diagram.

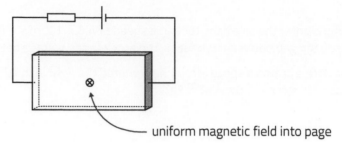

uniform magnetic field into page

Account for these different behaviours. Equations are not wanted. [QER 6]

(b) A tesla-meter consists of a probe containing a 'wafer' across which a Hall voltage is produced, and a 'meter unit' that supplies the wafer with a constant current, and also displays the measured value of the magnetic field.

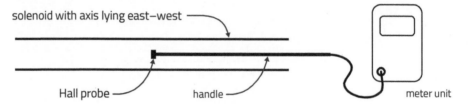

solenoid with axis lying east–west

Hall probe   handle   meter unit

Sally checks the calibration of the meter by placing its probe in the middle of a long solenoid with the probe orientated as shown, to produce the maximum reading.

The solenoid is 0.75 m long and has 150 turns of wire. Here are Sally's results:

solenoid current / A	0.0	0.2	0.4	0.6	0.8
meter reading / µT	3	53	104	154	204

(i) Suggest why Sally placed the solenoid with its axis lying East-West. [1]

(ii) Without drawing a graph, evaluate to what extent the results confirm that the tesla-meter is correctly calibrated. [3]

(c) Sally now uses the tesla-meter to investigate the magnetic field due to a long, straight, current-carrying wire. (See diagram.)

View from above

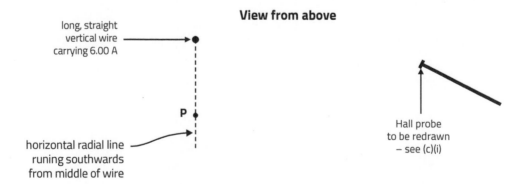

long, straight vertical wire carrying 6.00 A

P

horizontal radial line runing southwards from middle of wire

Hall probe to be redrawn – see (c)(i)

She places the probe at various points along the radial line, recording the meter reading and using a ruler to measure the distance, r, of the point from the centre of the wire.

(i) On the diagram redraw the Hall probe, positioned to measure the magnetic field at **P** due to the wire. [Note how the probe is positioned in part (a).] [1]

(ii) Sally's results are used to plot magnetic flux density, B, against $1/r$.

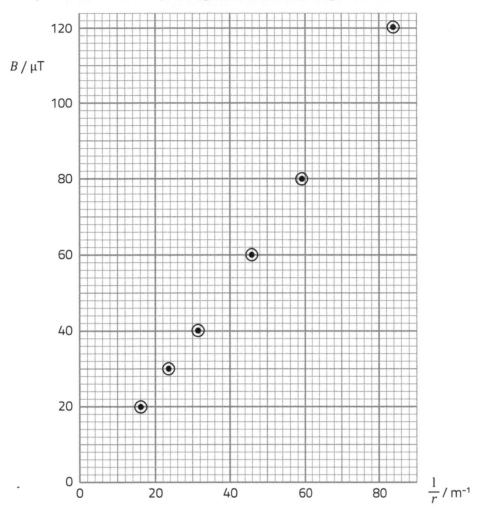

(I) Determine the current in the vertical wire. [4]

(II) Sally thinks that her values of r may be in error by a fixed amount because she has simply measured to the outer casing of the Hall probe. Her friend suggests that for this reason it would have been better to plot r against $\frac{1}{B}$. Discuss this suggestion. [2]

What is being asked

Part (a) is an AO1 question which combines two pieces of familiar bookwork. Even the diagram should be familiar. Hence AO1. The question looks for step-by-step explanation in a logical sequence with correct use of technical terms. Part (b)(i) is a straightforward AO2 mark but the mark allocation for (b)(ii) and 'to what extent' suggest there is more than one point to be made in this AO3 question.

Part (c)(i) tests knowledge of the shape of the wire's field with (c)(ii)I combining the analysis of a straight line with the expected equation for the wire's field – hence AO3. The last part is a more searching test of candidates' understanding of straight-line graph theory.

Mark scheme

Question part			Description	AOs			Total	Skills	
				1	2	3		M	P
(a)			**In a vacuum** • Magnetic force at right angles to velocity. • No increase in speed because no work done on particle or no force component parallel to velocity. • Force changes direction of particle velocity [at a constant rate]. **In conducting material** • Magnetic force deflects particles to the top (or bottom) of the slice. • Use of Left Hand Motor rule. • Displaced charged particles set up E field. • Force due to E opposes that due to B. • Equilibrium established when force due to E is equal (and opposite) to force due to B. • No resultant force so particles go straight.	6			6		
(b)	(i)		So Earth's magnetic field doesn't affect measurement, or equivalent. [1]		1		1		1
	(ii)		B calculated from $B = \mu_0 nI$ for at least one value of I, e.g. for 0.20 A, $B = 50.3$ mT [1] All readings correct **or** proportional to current [1] if zero error (or constant error) of 3 mT subtracted [1]			3	3	3	3
(c)	(i)		Probe at **P**; handle horizontal		1		1		
	(ii)	(I)	Straight line drawn close to points *except top one*, and through origin or within 1 small square to its right. [1] Gradient determined or equivalent. Accept poor line and/or slips in powers of 10 for this mark. [1] $B = \dfrac{\mu_0 I}{2\pi r}$ used, e.g. gradient $= \dfrac{\mu_0 I}{2\pi}$ [1] $I = 6.5$ [±0.1] A or 6.50 [±0.10] A **Unit** [1] [ecf on line drawn]			4	4	2	4
		(II)	$1/B$ against r would produce a straight line [despite error in r] but B against $1/r$ wouldn't. [1] Error in r appears as [−] intercept on r axis [or equiv.] [1]			2	2	2	2
Total				6	2	9	17	7	10

Rhodri's answers

(a) In a vacuum the force acts at right angles to the particle's speed, and makes it go in a circle. In the conducting material the same force makes the charged particles build up on the top of the slice and this stops the magnetic force from deflecting the particles any more, so they go straight.

MARKER NOTE
Apart from using 'speed' instead of the vector, 'velocity', Rhodri has correctly stated the direction of the force and realised that it's what makes the path a circle in a vacuum. He hasn't told us why the speed is constant. He seems to have some limited understanding of why the particles follow straight lines in the material. But he gives no reason for particles being deflected upwards, and he doesn't clearly say that their build-up is responsible for a second force opposing the magnetic force. A low-band answer.

2 marks

(b) (i) So that the Earth's magnetic field doesn't interfere with the results. ✓

MARKER NOTE
Correct answer.

1 mark

(ii) The reading is almost proportional to the current because 104 is almost twice 53 and so on. ✓ But the field should be exactly proportional to the current, so the meter calibration is good but not perfect.

MARKER NOTE
Middle mark gained for spotting the approximate proportionality, but analysis far from complete. Rhodri hasn't realised the need to compare a calculated value B for the solenoid with the results. Nor has he spotted the constant error.

1 mark

(c) (i) [Probe drawn at P with handle along broken radial line] ✗

MARKER NOTE
The probe is orientated to measure a radial B but the field is tangential.

0 marks

(ii) (I)

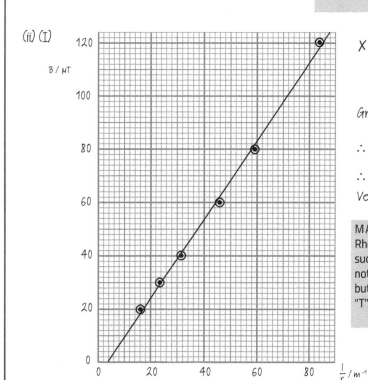

X

Gradient = $\dfrac{98}{67}$ = 1.4626 ✓

∴ 1.4626 = $\dfrac{4\pi \times 10^{-7} \times I}{2\pi}$ ✓

∴ $I = 7.313 \times 10^6$ A ✗
Very large current!

MARKER NOTE
Rhodri's line is quite close to all the points, but the three successive points below the line suggest that this was not the best choice. He then gains the middle two marks but loses the last because he missed the "μ" in front of "T" on the graph axis!

2 marks

(II) There will be no difference because $B = \dfrac{\mu_0 I}{2\pi r}$ and $r = \dfrac{\mu_0 I}{2\pi B}$ give lines of the same gradient. ✗

Except that an error in r will just shift the line up or down. ✓

MARKER NOTE
Rhodri's last sentence gains the second mark, but he hasn't made clear that, with zero errors in r, plotting B against $1/r$ doesn't give a straight line.

1 mark

| **Total** | **7 marks /17** |

Ffion's answers

(a) A charged particle in a magnetic field experiences a force at right angles to its velocity. This makes it change its direction continuously. It does this at a constant rate because the field is uniform. So it goes in a circle in a vacuum.

In the conducting material a Hall voltage builds up between top and bottom surfaces. This is because the particles are deflected upwards by the Left Hand Motor rule (if the particles are positive). The build-up of positive charge makes a downward force on the particles, opposing the force from the magnetic field. Soon the forces balance and the particles go straight (Newton's law 1).

MARKER NOTE
Ffion has accounted convincingly for the particle's circular motion in a vacuum, but not for the constancy of its speed.
She applied the Left Hand Motor rule correctly, and realised that the build-up of charge gives rise to a force that opposes the magnetic force. It would have been better still if she had mentioned the setting-up of an electric field, but her reference to the Hall effect and her correct use of Newton's first law confirm that this is a top band answer.

6 marks

(b)(i) The probe will not pick up the Earth's field, which is northwards. ✓

MARKER NOTE
Correct answer – very clear.

1 mark

(ii) For $I = 0.20$ A,
$B = 4\pi \times 10^{-7} \times (150/0.75) \times 0.20 = 50$ mT, ✓

and for the different currents used, $B = 0, 50, 101, 151, 201, 251$ mT. ✓

The readings are all 3 mT higher than these, so apart from a fixed error of 3 mT they are correct. ✓

MARKER NOTE
Although proportionality not mentioned, question answered fully.

3 marks

(c)(i) [Probe drawn at P with handle at right angles to radial line] ✓

MARKER NOTE
Probe correctly orientated.

1 mark

(ii) (I)

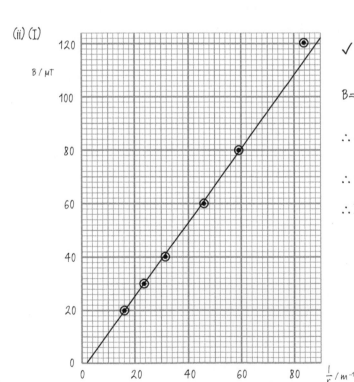

✓

$B = \dfrac{4\pi \times 10^{-7} I}{2\pi} \times \dfrac{1}{r}$

$\therefore$ Gradient $= \dfrac{4\pi \times 10^{-7} I}{2\pi}$ ✓

$\therefore \dfrac{120 \times 10^{-6}}{84 - 2} = \dfrac{4\pi \times 10^{-7} I}{2\pi}$ ✗

$\therefore I = 7.3$ A ✓[ecf]

MARKER NOTE
By ignoring the last point, Ffion's line fits the other points well. A good choice followed by competent analysis, though she has lost the last mark through reading 88 as 84 on the horizontal scale!

3 marks

(II) If there is an error of a in r. Then using r for the measured value, $r - a = \dfrac{\mu_0 I}{2\pi B}$.

$\therefore r = \dfrac{\mu_0 I}{2\pi B} + a$. So now the line will be straight ✓, with a as the intercept, when r is plotted against $1/B$. ✓

MARKER NOTE
Ffion has presented the theory of the proposed graph very clearly. Her use of the word 'now' implies that the original graph would not be straight.

2 marks

Total **16 marks /17**

Section 10: Electromagnetic induction

Topic summary

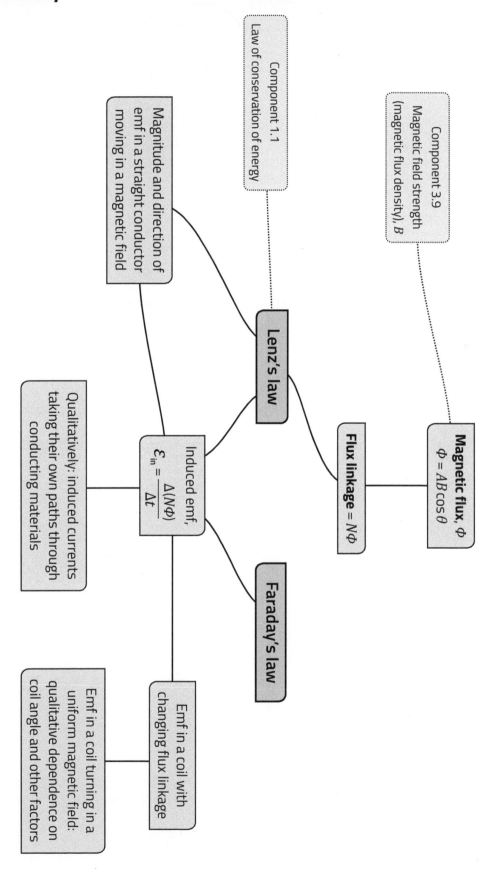

Component 1.1
Law of conservation of energy

Component 3.9
Magnetic field strength
(magnetic flux density), B

Magnitude and direction of
emf in a straight conductor
moving in a magnetic field

Lenz's law

Qualitatively: induced currents
taking their own paths through
conducting materials

Induced emf,
$\varepsilon_{in} = \dfrac{\Delta(N\Phi)}{\Delta t}$

Flux linkage $= N\Phi$

Magnetic flux, Φ
$\Phi = AB\cos\theta$

Faraday's law

Emf in a coil with
changing flux linkage

Emf in a coil turning in a
uniform magnetic field:
qualitative dependence on
coil angle and other factors

Q1 A ring made of copper wire has a **diameter** of 0.080 m. A uniform magnetic field of 0.050 T is applied to it, at right angles to the plane of the ring (into the paper in the diagram).

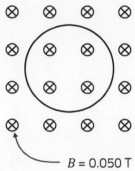

$B = 0.050$ T

(a) (i) Calculate the magnetic flux through the ring. [2]

....

....

....

(ii) Calculate the emf induced in the ring when the field strength is reduced to zero at a constant rate in a time of 0.16 s. [2]

....

....

....

(iii) State the sense of the induced emf (clockwise or anticlockwise in the diagram), justifying your answer clearly. [3]

....

....

....

....

....

(iv) The resistance of the ring is 2.75×10^{-3} Ω. Calculate the energy dissipated in the ring during the time that the field is changing. [2]

....

....

....

(b) Alice claims that doubling the diameter of the ring (but keeping the thickness of wire the same) would double the energy dissipated if the experiment of part (a) were repeated. [The magnetic field extends indefinitely.] Evaluate her claim. [2]

....

....

....

Q2   A square loop of wire, ABCD, measuring 0.15 m × 0.15 m is being pushed from left to right at a steady speed of 0.20 m s⁻¹.

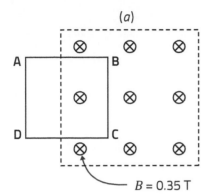
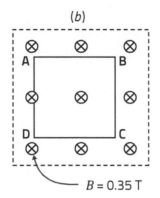
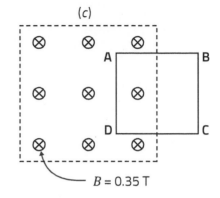

(a)   At the instant shown in diagram (*a*) the leading edge, **BC**, of the loop has entered a region of uniform magnetic field (bounded by the broken line). For this instant:

(i)   Calculate the emf induced in the square loop.   [2]

....

....

....

(ii)   Calculate the current, given that the loop resistance is 0.020 Ω, **and show the direction of the current on diagram (*a*)**.   [1]

....

....

(iii)   Calculate the motor effect (magnetic) force that acts on **BC and show its direction on diagram (*a*)**.   [3]

....

....

....

....

(b)   Explain why there is no current in the loop as it passes through the position shown in diagram (*b*).   [2]

....

....

....

(c)   **Draw an arrow**, correctly positioned on diagram (*c*), to show the force acting on the loop when the leading edge, **BC**, has passed out of the field.   [2]

Q3 (a) State Lenz's law of electromagnetic induction. [2]

....

....

....

(b) Using the set-up in question 2 (a) as your example, explain how Lenz's law is an application of Conservation of Energy. [2]

....

....

....

....

(c) Making use of your answers to question 2 (a), determine:

(i) The power dissipation by resistive heating in the loop. [2]

....

....

....

(ii) The work done per second pushing the loop into the field. [2]

....

....

....

....

Q4 A square coil of 150 turns measures 12.0 cm × 12.0 cm. It is placed so that the normal to the square is at an angle of 30° to the Earth's magnetic field at a place where its magnitude is 48 μT.

(a) Determine:

(i) The magnetic flux through the coil. [2]

....

....

....

(ii) The magnetic flux linkage. [1]

....

(b) The coil is now turned in a time of 1.2 s so that its normal is at right angles to the field. Calculate the mean emf induced in the coil. [2]

....

....

....

....

Q5 (a) State Faraday's law of electromagnetic induction. [2]

....

....

....

(b) The diagram is an edge-on view of a square coil, PQRS, which is being rotated at constant angular velocity, ω, about an axis passing through the midpoints of sides PQ and RS. The coil is in a uniform magnetic field, B, as shown. A resistor is connected across the coil.

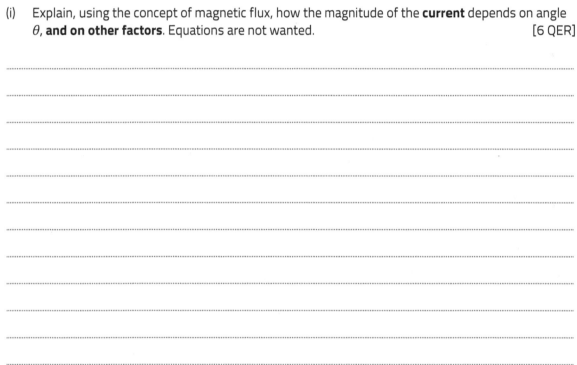

(i) Explain, using the concept of magnetic flux, how the magnitude of the **current** depends on angle θ, **and on other factors**. Equations are not wanted. [6 QER]

....

....

....

....

....

....

....

....

....

....

....

(ii) Sketch the variation of induced current with time over **two** complete revolutions of the coil. Start from the instant when $\theta = 0$. [2]

Q6

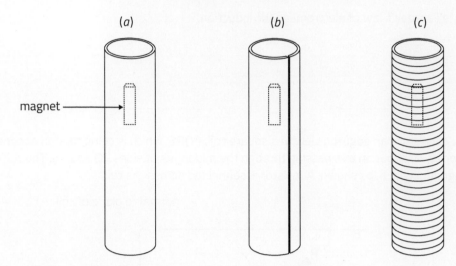

(a) A magnet is dropped down a vertical copper tube (diagram (a)). It does not touch the walls of the tube. The magnet approaches a terminal speed much sooner than it would if falling in a non-conducting tube of the same dimensions.

Explain carefully how the extra force resisting the motion arises. [3]

....

....

....

....

....

(b) A saw is used to make a narrow cut in the wall of the tube all down its length (diagram (b)). Explain briefly why the magnet now falls much more freely down the tube. [1]

....

....

(c) The tube in (a) is now cut into rings, and these are glued back together with insulating glue (diagram (c)). Nabila believes that the magnet's fall through the re-assembled tube will now be more like that in (a) than that in (b). Discuss whether or not her belief is likely to be correct. [2]

....

....

....

Q7 (a) A metal rod rests on metal rails a distance, l, apart, as shown. The circuit is completed by a resistor. There is a uniform magnetic field, B, directed into the paper.

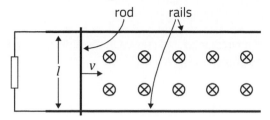

The rod is pushed to the right at a constant speed, v. A textbook gives two equations for the magnitude of the emf induced in this set-up:

$$E = Blv \quad \text{and} \quad E = \frac{\Delta \Phi}{\Delta t}$$

By considering the area swept out by the portion of rod between the rails in time Δt, show clearly that the two equations are equivalent. You may add to the diagram. [3]

....

....

....

....

(b) Starting at time $t = 0$, the rod shown in the left-hand diagram below is pushed to the right at a constant speed of 0.50 m s^{-1}. It makes contact with the conducting rails.

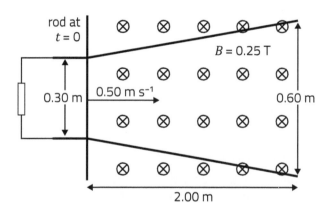

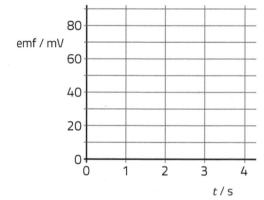

(i) Show that the emf initially is roughly 40 mV. [2]

....

....

(ii) **Use the graph grid** on the right (above) to show how the emf depends on time, t, over the interval $0 < t \le 4$ s. [2]

(iii) The resistance of the resistor is 1.50 Ω. The resistances of the rod and rails are negligible. Determine the current at $t = 3.0$ s. [2]

....

....

....

Question and mock answer analysis

Q&A 1

(a) A bar magnet is dropped through a flat circular coil, as shown in the side-view in Fig 1. The flux, Φ, through the coil varies with time, t, as shown in Fig 2.

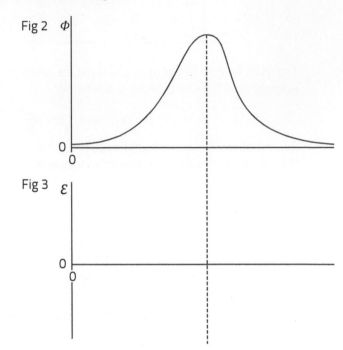

(i) Explain briefly why the flux falls more quickly than it rises. [1]

(ii) On the axes in Fig 3, sketch a graph of the emf induced in the coil against time. Take the initial emf as positive. [4]

(b) Tom connects a flat circular coil to a meter adapted to record the maximum emf, $\mathcal{E}_{max}$, induced in the coil when the magnet is dropped through it. He believes that:

$$\mathcal{E}_{max} = kv$$

in which k is a constant and v is the magnet's speed at the instant of maximum emf.

He drops the magnet four times from a height, h, above the coil (see Fig 1), recording $\mathcal{E}_{max}$ each time. He repeats the procedure for five more values of h, and plots points for $\mathcal{E}_{max}$ against h, together with their error bars on the graph grid. [See next page].

(i) Tom's values of $\mathcal{E}_{max}$ when $h = 0.400$ m are 6.0 mV, 6.1 mV, 6.3 mV, 6.1 mV.

Calculate the mean value and uncertainty in $\mathcal{E}_{max}^2$, and hence comment on whether the point and its error bar have been correctly plotted. [4]

(ii) Tom expects $\mathcal{E}_{max}$ and h to be related by the equation:

$$\mathcal{E}_{max}^2 = 2k^2g\,(h + h_0)$$

where h_0 is a small constant distance.

Justify this equation. [3]

(iii) Use the graph to find a value for k, together with its absolute uncertainty. [6]

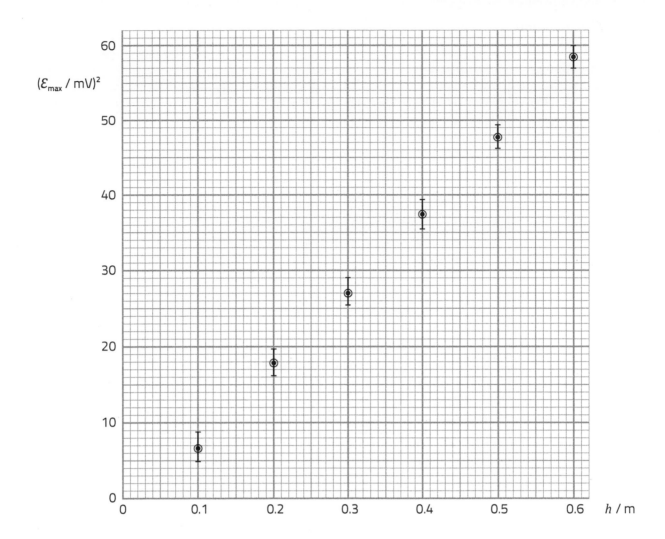

What is being asked

This is a fair question but sections (b)(ii) and (iii) are rather tough. This bit of the specification is conceptually demanding and this experiment is not one of the specified practicals. Having said that, (a)(i) is designed to be an easy AO2 mark, which draws the attention of the student to an important feature of the graph, whilst the mark allocation in (a)(ii) should alert the student to the fact that four features are expected. Part (b)(i) is a routine uncertainty question in which the presentation of results is important. Although the physics should be familiar in (b)(ii) the synoptic aspect adds to the difficulty. In (b)(iii) the procedure is not unique to the question but more steps are involved with no hints given, making this a rather testing AO3.

Mark scheme

Question part			Description	AOs 1	AOs 2	AOs 3	Total	Skills M	Skills P
(a)	(i)		Magnet leaves coil more quickly than enters. [1]		1		1		
	(ii)		Double hump with a zero at time of peak F. [1] Second hump inverted. [1] Second hump deeper than first is tall. [1] Second hump more compressed in time. [1]		4		4		
(b)	(i)		Mean of $\mathcal{E}_{max}^2$ = 37.5 or 38[(mV)^2 or × 10^{-6} V^2] [1] % unc in $\mathcal{E}_{max}^2$ = 2.4 [or 2.5 or 2.45 or by impl] [1] Unc in $\mathcal{E}_{max}^2$ = 1.8 or 1.9 or 2 [(mV)^2 or × 10^{-6} V^2] [1] Final presentation with units and consistent decimal places and with sensible comment (ecf) on plotted point, e.g.... 37.5 ± 1.8 (mV)^2 or 38 ± 2 × 10^{-6} V^2 so plotted point and error bar correct. [1]		4		4	4	4
	(ii)		h_0 linked to maximum emf not being when centre of magnet level with coil. [1] For next 2 marks accept omission or mishandling of h_0. $v^2 = u^2 + 2ax$ or $\frac{1}{2}mv^2 = mg(h[+h_0])$ leading to.... ... so $v^2 = 2g(h[+h_0])$ [1] Substitution from $\mathcal{E}_{max}^2 = kv$ so $\mathcal{E}_{max}^2 = 2k^2(h[+h_0])$ [1]	1 1	1		3	1 1	
	(iii)		At least one gradient evaluated using rise/run. Accept poor line and slips in power of 10 for this mark. [1] Max gradient between 107 × 10^{-6} [V^2 m^{-1}] and 111 × 10^{-6} [V^2 m^{-1}] [1] Min gradient between 95 × 10^{-6} [V^2 m^{-1}] and 99 × 10^{-6} [V^2 m^{-1}] [1] Penalise only once above for wrong power of 10. Correct use of gradient = 2 $k^2 g$. [1] k = 2.99 × 10^{-3} V m^{-1} s [or 3.0 × 10^{-3} V m^{-1} s] with ecf on gradients **UNIT** [1] ± 0.13 × 10^{-3} V m^{-1} s [or 0.1 or 0.2 × 10^{-3} V m^{-1}s] with ecf on gradients [1]			6	6	6	6
Total				2	10	6	18	12	10

Rhodri's answers

(a) (i) The magnet is falling faster as it leaves. ✓

(ii)

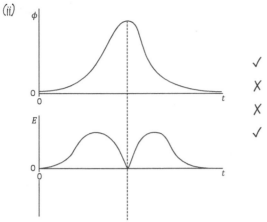

✓
X
X
✓

MARKER NOTE

Rhodri has grasped the essential point.

1 mark

MARKER NOTE

Rhodri realises that the magnitude of the emf has two maxima for the first mark and placed them suitably on the time scale for the second. He missed the points that the second maximum is greater (third mark) and that the emf is in the opposite sense when the magnet is receding (second mark).

2 marks

(b) (i) $\langle E_{max}^2 \rangle = \frac{1}{4}(36.0 + 37.21 + 39.69 + 37.21)$

$= 37.5$ ✓

$Unc = \frac{3.69}{2} =$ ✓✓ So plotting correct X

MARKER NOTE

Rhodri has chosen the simplest method for determining $\overline{\mathcal{E}_{max}^2}$ and its uncertainty and has carried it out correctly. except for omitting units and powers of 10. He is lucky that this mark scheme allows him...

3 marks

(ii) $(E_{max})^2 = 2k^2 g(h + h_0)^2$

$\therefore k^2 v^2 = 2k^2 g(h + h_0)^2$ ✓ [subst]

$\therefore v^2 = 2g(h + h_0)^2$

which is right for a falling body if h_0 is some error in measuring the drop height.

MARKER NOTE

Substituting from $\mathcal{E}_{max} = kv$ gains Rhodri the last mark, but his $v^2 = 2g(h + h_0)$ was obtained by working back from what he was supposed to show, and he doesn't know what h_0 represents.

1 mark

(iii)

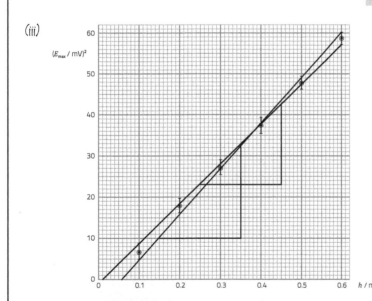

Max gradient $= \frac{33 - 10}{0.35 - 0.15}$ ✓

$= 115 \ mV^2 m^{-1}$ X

Min gradient $= \frac{42 - 23}{0.35 - 0.15}$

$= 95 \ mV^2 m^{-1}$ ✓

Comparing $(E_{max})^2 = 2k^2 g(h + h_0)^2$ with

$y = mx + c$, gradient $= 2k^2 g$

Max $k = \sqrt{\frac{115}{2 \times 9.81}} = 2.42$ ✓ X [unit]

min $k = \sqrt{\frac{95}{2 \times 9.81}} = 2.20$

$\therefore k = 2.3 \pm 0.1$ ✓ [ecf]

MARKER NOTE

Rhodri's steepest line just misses the error bar at 0.2 m. That, together with his rather small triangle has put the gradient outside the limits tolerated. However, his line of minimum gradient is within tolerance, so only one of the first three marks is lost. His method of determining a value for k and its uncertainty is valid, and his arithmetic is correct, except that he has forgotten, or given up on, units and powers of 10.

4 marks

| Total | 11 marks /18 |

Ffion's answers

(a)(i) The flux linking the coil is changing more quickly as the magnet leaves, because the magnet has gained speed. ✓

MARKER NOTE
Ffion has shown clear understanding.

1 mark

(ii)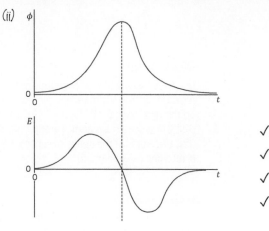

✓
✓
✓
✓

MARKER NOTE
Ffion's includes all four points mentioned in the mark scheme. Ideally, the peaks (positive and negative) should occur when the flux linkage is changing most rapidly (i.e. the points of inflexion) but the marking scheme does not demand that subtle point (the question is difficult enough!).

4 marks

(b)(i) $\overline{\varepsilon_{max}} = \frac{1}{4}(6.0 + 6.1 + 6.3 + 6.1)$ mV $= 6.13$ mV

$\therefore \overline{\varepsilon_{max}^2} = 38 \times 10^{-6}$ V² ✓

% uncertainty in $\varepsilon_{max} = \frac{0.15}{6.13} \times 100 = 2.45$ ✓

So % uncertainty in $\varepsilon_{max}^2 = 4.9$

So uncertainty in $\varepsilon_{max}^2 = 4.9 \times 37.6 \times 10^{-6}$ V²

$= 1.8 \times 10^{-6}$ V² ✓

Point and error bar correct. ✗ [d.p. error]

MARKER NOTE
Ffion's method is not the simplest but is perfectly valid. She loses the last mark because she gives $\overline{\varepsilon_{max}^2}$ to the nearest whole number × 10^{-6}, but the uncertainty to the nearest 0.1×10^{-6}.

3 marks

(ii) Max emf is when magnet is leaving coil and has fallen more than h.

If it has fallen by $(h + h_0)$ ✓

then by energy conservation

$\frac{1}{2}mv^2 = mg(h + h_0)$ ✓

But $\varepsilon_{max} = kv$ ✓ $\therefore \varepsilon_{max}^2 = 2k^2g(h + h_0)$

MARKER NOTE
The explanation is clear and correct. Ideally, the examiner would have liked to see the cancellation by m in the penultimate line but Ffion clearly knows what she is doing.

3 marks

(iii)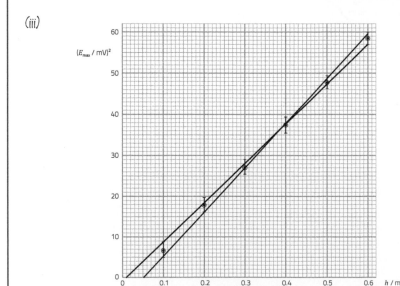

For steepest line, ✓✓

$Grad = \frac{60.0 - 0}{0.60 - 0.05} = 109 \times 10^{-6}$ V² m⁻¹

For least steep line, ✓

$Grad = \frac{57.0 - 0}{0.600 - 0.010} = 97 \times 10^{-6}$ V² m⁻¹

So gradient $= 103 \pm 6 \times 10^{-6}$ V² m⁻¹

So $k = \sqrt{\frac{grad}{2g}}$ ✓ $= 2.29 \frac{mV}{m\,s^{-1}}$ ✓

Uncertainty $= 2.3 \times \frac{6}{103} \frac{mV}{m\,s^{-1}}$

$= 0.13 \frac{mV}{m\,s^{-1}}$ ✓

MARKER NOTE
Ffion's lines are the steepest and least steep allowed by the error bars and, although she has not drawn triangles, her gradient calculations show that she has used the largest possible. She has calculated k and its uncertainty correctly, and their units are correct though not given in the standard format. This really is an excellent answer from a candidate who knows what she is about.

6 marks

Total **17 marks /18**

Option A: Alternating currents

Topic summary

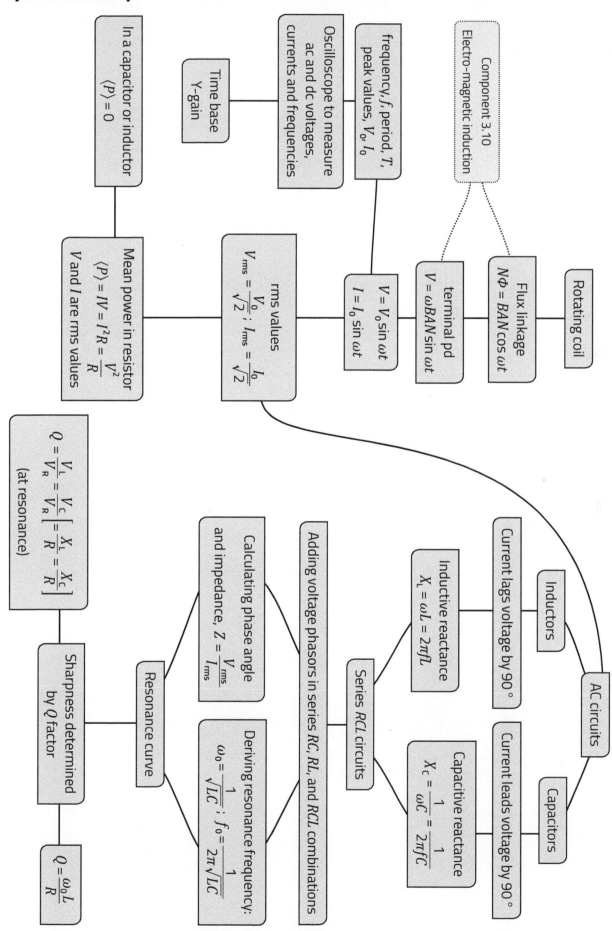

In a capacitor or inductor
$\langle P \rangle = 0$

Time base
Y-gain

Oscilloscope to measure
ac and dc voltages,
currents and frequencies

frequency, f, period, T,
peak values, V_0, I_0

Component 3.10
Electro-magnetic induction

Mean power in resistor
$\langle P \rangle = IV = I^2 R = \dfrac{V^2}{R}$
V and I are rms values

rms values
$V_{\text{rms}} = \dfrac{V_0}{\sqrt{2}}$; $I_{\text{rms}} = \dfrac{I_0}{\sqrt{2}}$

$V = V_0 \sin \omega t$
$I = I_0 \sin \omega t$

terminal pd
$V = \omega BAN \sin \omega t$

Flux linkage
$N\Phi = BAN \cos \omega t$

Rotating coil

$Q = \dfrac{V_L}{V_R} = \dfrac{V_C}{V_R} \left[= \dfrac{X_L}{R} = \dfrac{X_C}{R} \right]$
(at resonance)

Calculating phase angle
and impedance, $Z = \dfrac{V_{\text{rms}}}{I_{\text{rms}}}$

Adding voltage phasors in series RC, RL, and RCL combinations

Series RCL circuits

Inductive reactance
$X_L = \omega L = 2\pi f L$

Current lags voltage by $90°$

Inductors

AC circuits

Sharpness determined
by Q factor

Resonance curve

Deriving resonance frequency:
$\omega_0 = \dfrac{1}{\sqrt{LC}}$; $f_0 = \dfrac{1}{2\pi\sqrt{LC}}$

Capacitive reactance
$X_C = \dfrac{1}{\omega C} = \dfrac{1}{2\pi f C}$

Current leads voltage by $90°$

Capacitors

$Q = \dfrac{\omega_0 L}{R}$

Q1 A square coil, PQRS, of 50 turns, measuring 4.00 cm × 4.00 cm, is turning in a magnetic field of 0.150 T at 20.0 revolutions per second, about an axis passing through the midpoints of PQ and RS. The diagram shows the coil, edge-on, at time $t = 0$.

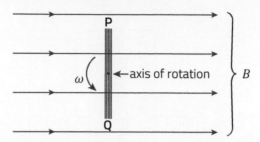

(a) (i) State one time at which the flux linkage is a maximum. .... [1]

    (ii) Calculate the maximum flux linkage. [2]

(b) (i) State one time at which the induced emf is a maximum. .... [1]

    (ii) Calculate the maximum emf. [2]

(c) (i) Calculate, to 3 significant figures, the *change* in flux linkage occurring between $t = 0.0115$ s and $t = 0.0135$ s. [3]

    (ii) **Hence** calculate the mean emf induced in the coil during this time interval. [2]

    (iii) Compare your answer to (c) (ii) with the instantaneous emf at $t = 0.0125$ s and explain whether or not the comparison is as expected. [3]

Q2 (a) The coil of a simple ac generator has a resistance of 2.4 Ω and is connected (by slip rings and brushes of negligible resistance) to a 5.6 Ω 'load' resistor. The power dissipated in the **resistor** is 0.30 W.

(i) Calculate the rms pd across the **resistor**. [2]

....

....

....

(ii) Show that the emf of the generator is roughly 2 V rms. [Hint: the resistance of its coil is its *internal resistance*.] [2]

....

....

....

(b) Calculate the rotation frequency (number of revolutions per second) needed to generate this rms emf. The coil is square, measures 5.0 cm × 5.0 cm, consists of 100 turns and rotates in a uniform magnetic field of 0.30 T. [3]

....

....

....

....

....

Q3 A resistor, a capacitor and an inductor are connected in series across a signal generator which is set to 500 Hz. The rms pds across the components are shown in the diagram.

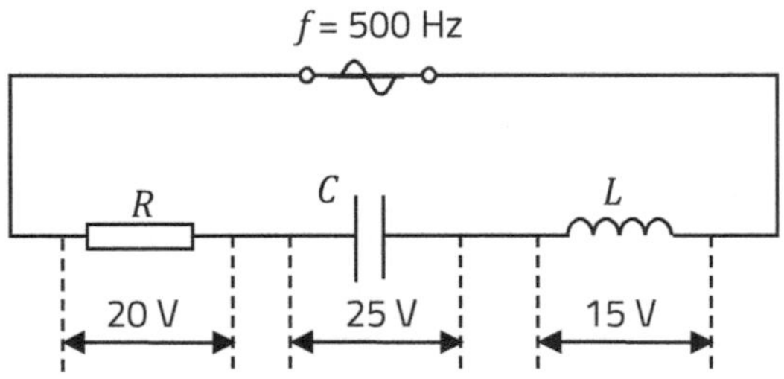

(a) Draw a voltage phasor diagram and use it to determine the rms pd across the terminals of the signal generator. [3]

$f = 500$ Hz

R C L

20 V   25 V   15 V

....

....

....

....

....

(b) Ciaran says that the resonance frequency of the circuit is greater than 500 Hz. Discuss briefly whether he is correct. [2]

....

....

....

Q4 An alternating pd is applied across the Y-input terminals of an oscilloscope with its Y-gain set to 200 mV / div and its time base to 2 ms / div. The display is shown:

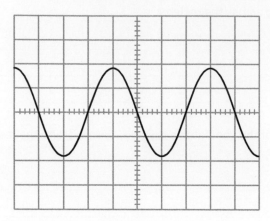

(a) Determine the frequency of the pd. [2]

....

....

....

(b) (i) Determine the rms value of the pd **and** give a reasoned estimate of its absolute uncertainty. [4]

....

....

....

....

....

(ii) The following Y-gains were also available: 50 mV / div, 100 mV / div, 500 mV / div. Evaluate whether or not 200 mV / div was the best Y-gain to have chosen for examining this pd. [2]

....

....

....

Q5 A sinusoidal alternating current of peak value 0.15 A is in a series combination of a 12 Ω resistor and a capacitor of reactance 20 Ω. Calculate:

(a) the rms pd across each component; [2]

(b) the mean power dissipated. [2]

Q6 The graph shows the pd, V, applied across a capacitor of capacitance 0.60 mF:

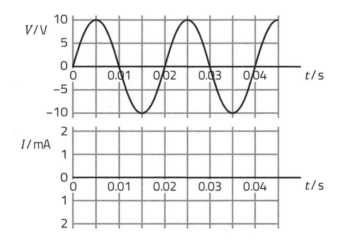

(a) Show that the peak current is approximately 2 mA. [3]

(b) **Sketch a graph** of current against time on the grid provided. [2]

(c) Basing your answer on the definitions of *current* and *capacitance*, explain why the maxima of current occur at the times where you have shown them. [3]

Q7 The coil of a small generator has 240 turns and an area of 20 cm². It rotates at 1500 revolutions per minute at right angles to a uniform field of 45 mT. Assuming that the coil has negligible resistance, calculate the mean power it delivers to a load of resistance 120 Ω. [4]

Q8 (a) State one way in which *reactance* is similar to *resistance*. [1]

(b) State one *way* in which *reactance* differs from *resistance*. [1]

Q9 The reactances of a capacitor and an inductor at 50 Hz are shown as points marked X_c and X_L on the grid:

(a) **Add lines**, straight or curved as appropriate, to the grid to show how these reactances vary with frequency. [3]

(b) Calculate:

(i) the capacitance of the capacitor; [2]

(ii)   the inductance of the inductor.   [2]

....

....

....

Q10   An inductor has a reactance of 15 Ω at 0.5 kHz. This is indicated on the grid.

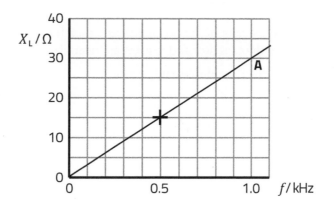

(a)   State the significance of graph **A** on the grid.   [1]

....

....

(b)   The inductor is connected in series with a resistor of resistance 10 Ω. Sketch a second graph to show the variation of the impedance of the combination between 0 and 1.0 kHz.   [3]

Space for calculations:

Q11 The diagram shows a capacitor and a resistor connected in series across a sinusoidal power supply:

10.0 V$_{rms}$ at 750 Hz

0.47 μF

330 Ω

(a) (i) In the space to the right of the circuit diagram, sketch a labelled phasor diagram to show how the rms pds across the capacitor and resistor are related to the rms pd of the supply. [2]

(ii) Show that the rms current is roughly 20 mA. [3]

(b) (i) Calculate the rms pd, V_c, across the capacitor. [2]

(ii) Determine the phase angle between V_c and the pd applied across the RC combination. [2]

(iii) Andrew claims that the rms pd across the resistor is (10.0 V − V_c). Evaluate his claim. [2]

(c) Calculate:

(i) the energy dissipated in the circuit **per cycle**; [2]

(ii) the mean energy stored in the capacitor. [2]

Q12 A coil of wire behaves as an inductance, L, in series with a resistance, R. A group of students uses the circuit shown to investigate how the impedance of a coil varies with frequency.

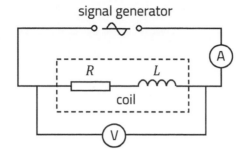

(a) Describe briefly how the circuit can be used to determine the *impedance* of the coil at 100 Hz. The meters are calibrated to read rms values. [2]

....

....

....

(b) With the aid of a labelled phasor diagram, derive the equation:

$$Z^2 = 4\pi^2 L^2 f^2 + R^2$$

[You may use the equation $X_L = \omega L$.] [2]

....

....

....

....

....

....

(c) The students plot a graph of Z^2 against f^2:

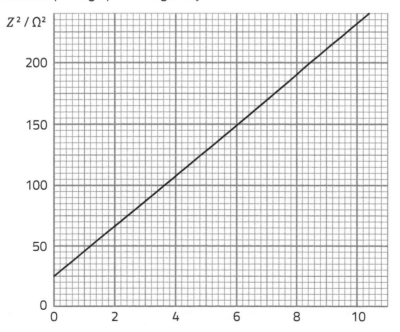

Determine:

    (i)   the coil's resistance;   [2]

    (ii)   the coil's inductance.   [3]

Q13 (a) The values for the resistance, R, and inductance, L, of the coil in question 12 are checked by a different method. A 0.100 μF capacitor is connected in series with the coil in the circuit of 12(a). The signal generator output is kept at 5.00 V rms and the frequency varied. A maximum current of 1.00 A rms is found to occur at 10.6 kHz.

Giving your working, use this information to calculate values for R and L.   [3]

  (b)   Calculate:

    (i)   the rms pd across the capacitor at 10.6 kHz;   [2]

    (ii)   the Q factor of the circuit.   [1]

Q14   A student reads that the inductance of an air-cored solenoid is given by $L = \mu_0 \dfrac{N^2 A}{\ell}$,where N is the number of turns, A the cross-sectional area and ℓ the length. She winds a 15 mm long, 3.0 mm radius inductor with 25 turns of wire. Calculate the capacitance of the capacitor she needs to connect in series to produce a circuit of resonance frequency 1.6 MHz.   [3]

Q15 (a) A coil of wire of resistance 33 Ω and inductance 0.070 H is connected to a signal generator set to 100 Hz.

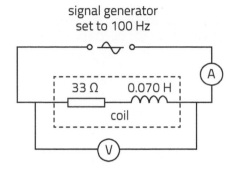

signal generator
set to 100 Hz

The rms current through the coil is 0.20 A. Giving a labelled phasor diagram, show that the rms pd across the signal generator's terminals is roughly 10 V. [5]

....

....

....

....

....

....

(b) (i) Calculate the value of capacitor that must be included in the circuit, in series with the coil, in order to produce resonance at that frequency. [2]

....

....

....

(ii) Calculate the new rms current, assuming that the rms pd across the signal generator's terminals is unchanged. [2]

....

....

....

(c) If the frequency of the signal generator is varied, keeping its rms pd constant, a resonance curve (current against frequency) can be plotted for the set-up in (b). Ciaran claims that the curve will be identical if the experiment is repeated with the 33 Ω resistor replaced by a 66 Ω resistor and the pd of the signal generator doubled. Evaluate to what extent he is correct. [3]

....

....

....

....

....

....

Question and mock answer analysis

Q&A 1   This question is about the circuit shown below. The signal generator is set to give an output of 6.00 V rms at any frequency selected.

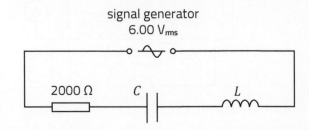

(a)   It is known that at 65 Hz the reactance of the capacitor is 8.40 kΩ and the reactance of the inductor is 2.10 kΩ.

(i)   Calculate the inductance of the inductor.   [2]

(ii)   (I)   Sketch a labelled phasor diagram of voltages at 65 Hz, and show that the rms current is approximately 1 mA.   [5]

(II)   Explain why the rms current has the same value at 260 Hz.   [2]

(iii)   A grid for a resonance curve (rms current against frequency) is given.

(I)   By making appropriate calculations, show that the point already plotted is at the correct position for the **top** of the curve.   [3]

I / mA

3

2

1

0

0   100   200   300   400 f / Hz

(II)   Use (a)(ii)(I) and (II) to plot two more points, and sketch the resonance curve between $f = 0$ and $f = 400$ Hz.   [2]

(iv)   Calculate the Q factor for the circuit.   [2]

(b)   The inductor is now replaced by one of inductance $\frac{1}{2}L$, and the capacitor by one of capacitance $2C$. Emma claims that the resonance frequency and the current at resonance will stay the same, but the curve will be sharper. Evaluate her claims.   [4]

What is being asked

The question opens in a non-frightening way, with (a)(i) just involving the use of a standard equation. In part (ii) the examiner wishes to test understanding of phasor diagrams and has saved candidates the trouble of working out reactances from L and C values! There is more than one way of proceeding in the second part, but the low mark allocation suggests that this part can be done simply. Although candidates are being asked about a point on a graph in (iii)(I), very standard calculations are being tested but part (II) requires recall of knowledge.

(a)(iv) is a straightforward test of the candidate's ability to calculate a specific quantity, but perhaps its inclusion at this point could be a clue to help with the next part ... (b) in which the candidate has to evaluate three claims, choosing the order in which to tackle them, and the strategy – hence the AO3 classification.

Mark scheme

Question part			Description	AOs			Total	Skills	
				1	2	3		M	P
(a)	(i)		$L = \dfrac{X_L}{\omega}$ (transposition at any stage) [1] $L = 5.1$ H [1]		2		2	2	
	(ii)	(I)	Pattern of 3 lines or arrows labelled to identify as X_L, R, X_C or V_L, V_R, V_C even if X_L, X_C or V_L, V_C wrong way round. [1] X_L (or V_L) shown π in advance of X_C (or V_C) [1] $Z = \sqrt{(8400 - 2100)^2 + 2000^2}$ equiv or by impl [1] $I = V / Z$ used [1] $I = 0.91$ mA [1]	1 1 1 1	1		5	3	
		(II)	Values of X_L and X_C swap over or equivalent [1] $(8400 - 2100)^2 = (2100 - 8400)^2$ or equiv [1]		1 1		2	2	
	(iii)	(I)	X_L shown to equal X_C e.g. both stated to be 4.2 kΩ, at 130 Hz [1] $I = 6.00$ [V] / 2000 [Ω] = 3.0 mA as plotted [1] Resistance only, or cancelling of reactances, shows current has greatest value. [1]		1 1 1		3		1
		(II)	Points at 65 Hz and 260 Hz plotted correctly and bell-shaped curve drawn through the 3 points [1] Curve goes through origin and looks as if it *could* be asymptotic to f axis for large f. [1]	1	1		2		
	(iv)		Use of $X = 4.2$ kΩ [1] $Q = 2.1$ [1]	2			2		
(b)			$\omega_0^2 = \dfrac{1}{\frac{1}{2}L2C} = \dfrac{1}{LC}$, or equiv argument [1] [V (rms) and] R the same so I same at res [1] X_L at resonance halved because ω_0 the same and L halved or equivalent argument [1] Therefore Q halved **and** curve *less* sharp so Emma incorrect. [1]			4	4		
Total				7	9	4	20	8	0

Rhodri's answers

(a) (i) $L = \dfrac{2100}{65} = 32$ H

(ii)(I)

8400

2000

2100

✓ X

$Z = \sqrt{(8400 - 2100)^2 + 2000^2}$ ✓

$= 5974$ X

$I = \dfrac{6.00}{5974}$ ✓ ecf

$= 1.00 \times 10^{-3}$ A X

$= 1.00$ m

(II) Because the current reaches a peak and comes back down, there must be another frequency at which the current has the same value as at 65 Hz.

(iii) (I) At top of curve, Z has lowest value possible ✓
so Z = R, so I = 6.00 / 2000 = 3 × 10⁻³ A = 3 mA, so
point is correctly plotted ✓

(II).

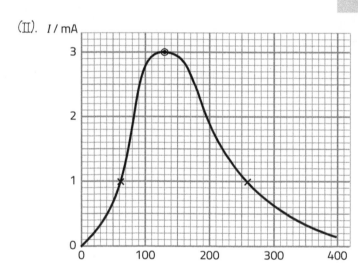

(iv) $Q = \dfrac{X_L}{R} = \dfrac{2100}{2000} = 1.1$ X

(b) At resonance $\omega_0 L = \dfrac{1}{\omega_0 C}$ ∴ $\omega_0^2 = \dfrac{1}{LC} = \dfrac{1}{(\frac{1}{2}L)\, 2C}$ ✓
Therefore resonance frequency the same.
Resistance the same so current the same. ✓

Sharpness depends on resistance, so sharpness the same.
Emma has got this one wrong.

Ffion's answers

(a)(i) $X_L = \omega L$

$\therefore L = \dfrac{X_L}{\omega}$ ✓ $= \dfrac{2100}{2\pi \times 65} = 16$ H ✗

> **MARKER NOTE**
> Ffion's transposition and substitutions are correct, but her numerical calculation is wrong (she has omitted the π). She loses only the second mark.
> **1 mark**

(ii)(I)

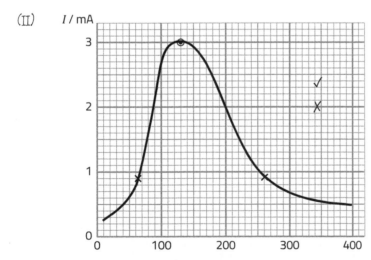

$V = \sqrt{I^2(X_L - X_C)^2 + I^2 R^2}$

$I = \dfrac{V}{\sqrt{(X_L - X_c)^2 + R^2}}$ ✓

$= \dfrac{6.00}{\sqrt{(2100 - 8400)^2 + 2000^2}}$ ✓ $= 0.908$ mA ✓

> **MARKER NOTE**
> Ffion's first phasor diagram is excellent. Her second, though useful, isn't strictly needed in order to answer the question as set. Her calculation is correct.
> **5 marks**

(II) $X_c \propto \dfrac{1}{f}$ and $260 = 4 \times 65$

$\therefore$ at 260 Hz, $X_c = \frac{1}{4}$ 8.4 kW = 2.1 kW

Also X_L μ f $\therefore$ at 260 Hz, $X_L = 4 \times 2.1$ kW

So X_L and X_c are simply exchanged. ✓

> **MARKER NOTE**
> Ffion has thoroughly earned the first mark, but she has not explained why the exchange of X_c and X_L leaves Z the same.
> **1 mark**

(iii)(I) At 130 Hz, $X_c = 4.2$ kW, $X_L = 4.2$ kW ✓
Reactances cancel so $Z = R$ and is a minimum.
Therefore $I = 6.00/2000 = 3.00$ mA ✓ and is
a maximum at 130 Hz. ✓

> **MARKER NOTE**
> The explanation is clear and correct. The 'maximum' mark is earned because of the statement on the second line.
> **3 marks**

(II)

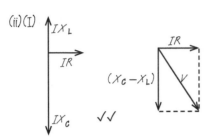

> **MARKER NOTE**
> Ffion has drawn a bell-shaped curve through the points at 65 Hz, 130 Hz and 260 Hz, but it will not go through the origin, as it must because X_c is unbounded as f approaches zero.
> **1 mark**

(iv) $Q = \dfrac{X_L}{R} = \dfrac{2100}{2000}$ ✓ $= 2.1$ ✓

> **MARKER NOTE**
> Q is correctly calculated.
> **2 marks**

(b) $X_c = \dfrac{1}{C}$ so doubling C halves X_c for any frequency.
$X_L = \omega L$ so halving L halves X_L. Therefore we still have
$X_c = X_L$ at 130 Hz. ✓

Resistance unchanged so current still 3.0 mA ✓

Q is decreased, so curve is now
less sharp – and Emma is wrong!
[not enough]

> **MARKER NOTE**
> Ffion has shown (perhaps not in the most direct way) that Emma's first two claims are correct. She is correctly attributing less sharpness to a lower Q, but she has not clearly shown Q to be lower – hence the third mark lost.
> **2 marks**

Total	15 marks /20

Option B: Medical physics

Topic Summary

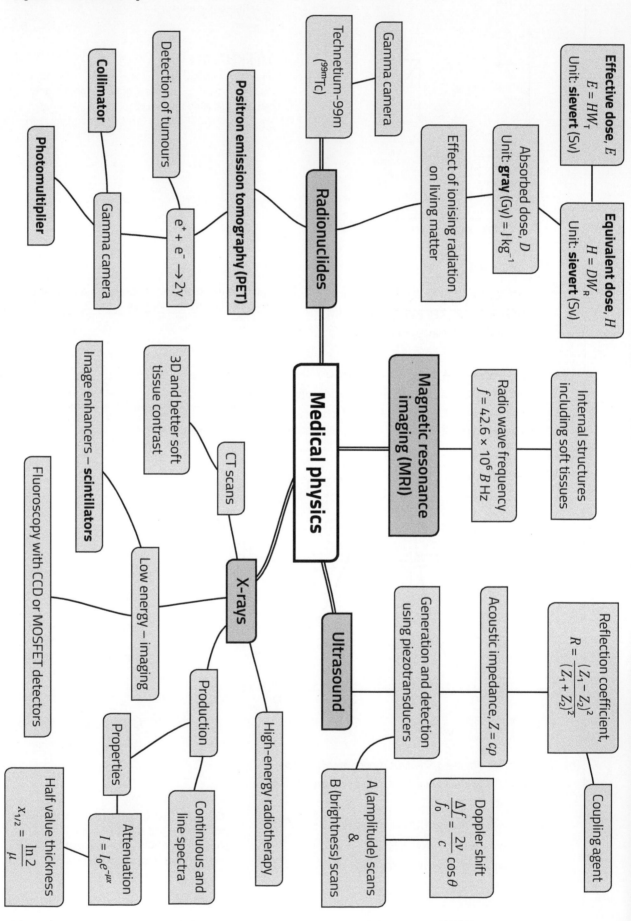

Q1 State some of the properties of X-rays that make them suitable for imaging bones. [3]

Q2 An X-ray tube is shown:

(a) Explain the process by which the continuous spectrum is produced. [2]

(b) Explain the process by which the line spectrum is produced. [2]

(c) Explain why a vacuum is required. [1]

(d) Calculate the speed of the electrons just before they strike the tungsten target and comment on the validity of your calculation. [3]

(e) Calculate the minimum wavelength of X-rays emitted. [2]

(f) The tube current is 14.5 mA and 5.1 W of X-ray power is produced.

(i) Calculate the efficiency of the X-ray tube. [2]

(ii) Explain why water-cooling of the tungsten target is required. [2]

(iii) A book on X-rays states that a good approximation for the efficiency of an X-ray tube is:

Efficiency of X-ray tube (%) = pd (in kV) × atomic number of target × 10^{-4}

Determine whether this equation is a good approximation for this X-ray tube. The atomic number of tungsten is 74. [2]

Q3 The X-ray spectrum of an X-ray tube operating at 60 kV is shown.
On the same diagram, sketch the X-ray spectrum for the same tube operating at 30 kV. [3]

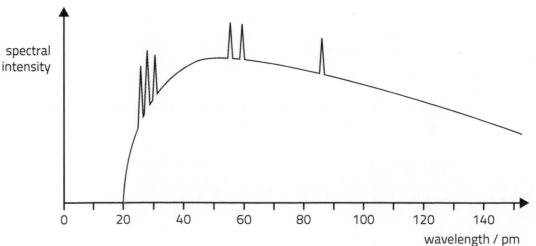

Q4 A book on medical physics states:

A piezoelectric transducer emits ultrasound by using the reverse piezoelectric effect while a piezoelectric detector uses the piezoelectric effect.

(a) (i) Explain the difference between the *reverse piezoelectric effect* and the *piezoelectric effect*. [2]

(ii) Explain how a piezoelectric transducer can be used to send short ultrasound pulses. [2]

(b) Explain the difference between an ultrasound A-scan and an ultrasound B-scan. [4]

(c) The table shows values of density, speed of sound and acoustic impedance for soft tissue and bone:

Body tissue	Density / kg m^{-3}	Speed of sound / m s^{-1}	Acoustic impedance / ....
Soft tissue		1590	1.70×10^6
Bone	1650		6.73×10^6

(i) Complete the table including the correct unit for acoustic impedance. [3]

(ii) Show that the reflection coefficient between soft tissue and bone is 36%. [1]

(iii) A new-born baby's brain can be imaged using an ultrasound scan because there is no calcification of the skull. A radiologist states that the detected reflections from inside the brain will be many times weaker for an adult compared with that of a new-born child. She further states that the intensity of reflections will be 13% of those for the baby because $0.36^2 = 0.13$, for the reflection on the way in and on the way out. Determine to what extent the radiologist is right and correct her figures (if they need correcting). [3]

Q5 (a) The half thickness $(x_{\frac{1}{2}})$ is defined as the thickness of a material that reduces the intensity of X-rays to half its original value. Show that the half thickness is related to the absorption coefficient (μ) by the following relationship:

$$\mu x_{\frac{1}{2}} = \ln 2$$

[3]

(b) In muscle fibre, the half thickness for certain X-rays is 3.7 cm. Calculate the percentage reduction in X-ray intensity when the beam of X-rays passes through 5.0 cm of muscle fibre. [3]

Q6 (a) An ultrasound technician takes images of a foetus inside the womb using an ultrasound B-scan. Explain briefly how an ultrasound B-scan image is produced and why a coupling gel is crucial in obtaining the image. [4]

(b) (i) As part of the ultrasound scan, the maximum speed of blood flow in the aorta of the foetus is measured. Explain how the blood flow can be measured using the Doppler shift. [3]

....

....

....

....

....

....

(ii) Ultrasound of frequency 5.50 MHz is used, the ultrasound enters the aorta at an angle of 5.0° to the direction of blood flow and the speed of sound in blood is 1580 m s^{-1}. Calculate the change in frequency of the ultrasound when the blood flow is measured as 105.8 cm s^{-1}. [2]

....

....

....

....

Q7 A simplified diagram of a fluoroscopy set-up is shown.

(a) Explain the purpose of the anti-scatter grid and the scintillator screen. [2]

....

....

....

....

....

....

CCD video camera

scintillator screen

anti-scatter grid

patient

X-ray filter

simple collimator

X-rays

X-ray tube

(b) Explain why the design of the scintillator screen is important in decreasing the risk of cancer to the patient. [2]

....

....

....

....

....

Q8 High energy X-rays (around 10 MeV) are used in radiation therapy.

(a) Explain why higher energy X-rays are required for radiation therapy than for X-ray imaging. [2]

(b) Explain why higher intensity beams are required for radiation therapy. [2]

Q9 (a) Explain the role of radio waves in magnetic resonance imaging (MRI). [3]

(b) Calculate the range of frequency of radio waves used by an MRI machine which has a magnetic field strength varying from 1.53 T to 1.94 T. [2]

(c) An international netball player has a problem with her knee joint. An orthopaedic surgeon recommends an MRI scan, traditional X-ray images and an ultrasound B-scan to diagnose the problem. Evaluate the strengths and weaknesses of each of these techniques in imaging the knee joint. [6]

Q10 (a) In a PET scan, two gamma rays are detected by two gamma detectors **A** and **B**. The gamma ray is detected by detector **A** 237 ps before the other gamma ray is detected by detector **B**. Place a cross on the diagram at the location of the source of the two gamma rays. [3]

detector A   detector B

0cm1 2 3 4 5 6 7 8 9 10 11 12 13 14 15 16 17 18 19 20 21 22 23 24 25 26 27 28 29 30

[Space for calculations]

(b) Explain briefly the physical principles of a PET scan. [4]

Q11 Explain briefly how a CT scanner can obtain a 3D image of a patient's innards. [4]

Q12 Tables of the radiation weighting factor and tissue weighting factor are shown:

Radiation type and energy range	Radiation weighting factor W_R
X-rays and γ-rays, all energies	1
Electrons, positrons, muons, all energies	1
Neutrons:	
<10 keV	5
10 keV to 100 keV	10
>100keV to 2 MeV	20
>2 MeV to 20 MeV	10
>20 MeV	5
Protons	2 to 5
α-particles	20

Tissue	Tissue weighting factor W_T
Red bone-marrow, colon, lung, stomach, breast, other tissues	0.12
Gonads	0.08
Bladder, oesophagus, liver, thyroid	0.04
Bone, brain, salivary glands, skin	0.01

(a) A patient is irradiated uniformly with a beam of 1 MeV neutrons. The equivalent dose received by the liver is 550 mSv.

(i) Calculate the contribution of the liver to the patient's effective dose. [2]

(ii) The neutrons are absorbed uniformly by the patient's whole body and the patient has a mass of 94 kg. Calculate the total energy of the neutrons absorbed. [3]

(b) A radiologist states that breathing or swallowing an α-particle source may be the most dangerous thing you could possibly do with ionising radiation. Use data from the tables to evaluate this statement. [3]

Q13 The diagram shows a patient who has received the radioactive tracer technetium-99m. A gamma camera is then used to image the patient's kidneys.

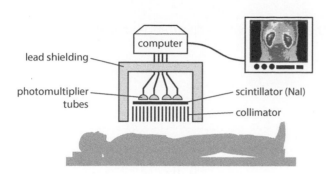

(a) Explain the purpose of the scintillator, collimator and photomultiplier tubes in the gamma camera. [3]

....

....

....

....

....

(b) State some properties you would expect of technetium-99m for it to be a suitable radionuclide to be used with the gamma camera. [3]

....

....

....

....

....

....

Question and mock answer analysis

Q&A 1

(a) Calculate the operating pd of the X-ray tube whose spectrum is shown. [3]

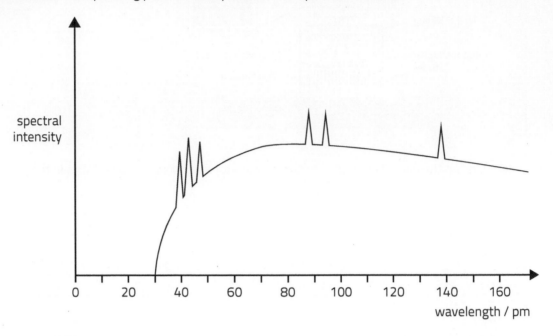

(b) On the diagram above, sketch the X-ray spectrum for the same tube when the operating pd is increased by 50%. [3]

(c) Five years after an operation for a replacement heart valve, a patient is required to have an annual scan to check the performance of the valve. Evaluate the suitability of the following techniques in carrying out the annual scan:

Fluoroscopy   CT   MRI   Ultrasound   PET   [5]

What the question is asking

Part (a) are AO2 marks. They require the candidate to obtain a value for the minimum wavelength from the graph and then use this to obtain the accelerating pd of the X-ray tube.

Part (b) are mainly AO1 marks but there is one AO2 mark for calculating the new minimum wavelength. The two AO1 marks are for knowing that the basic shape of the spectrum will remain the same and for knowing that the position of the spikes must remain at the same wavelength.

Part (c) are all AO3 marks. The candidates must evaluate the suitability of each of the five imaging techniques in obtaining information about a heart valve. The strengths and weaknesses of each technique are required for a good answer.

Mark scheme

Question part			Description	AOs			Total	Skills	
				1	2	3		M	P
(a)			Minimum wavelength read correctly (30 pm) [1] $eV = hc/\lambda$ applied [1] Correct answer = 41.4 kV [1]		3		3	1	
(b)			Minimum wavelength = 20 pm (written or in graph) [1] All peaks in same place [1] Background spectrum following same pattern [1]	1 1	1		3	1	
(c)			Any 2 points for each method – 1 mark **Fluoroscopy** – moving images, flow can be checked from videos, X-ray dosage, expensive. **CT** – image only (not moving), X-ray dosage, blood flow can be checked (using contrast media), expense **MRI** – good images, expensive, non-ionising, blood flow can be checked (using contrast/real time MRI) **Ultrasound** – B-scan cheap, moving images, non-ionising, Doppler gives blood flow too (best method) **PET** – radiation dosage, low resolution, still images, not useful for flow (expensive or availability not a mark option because it will not work)			5	5		
Total				2	4	5	11	2	0

Rhodri's answers

(a) $V = \dfrac{hc}{e\lambda}$ ✓

$= \dfrac{6.63 \times 10^{-34} \times 3 \times 10^{8}}{1.6 \times 10^{-19} \times 30 \times 10^{-9}}$ ✓ $= 41.4 \text{ V}$ ✗

MARKER NOTE
Rhodri's answer is perfect except that he has made a mistake with the power of ten (pm is 10^{-12} m not 10^{-9} m).

2 marks

(b)

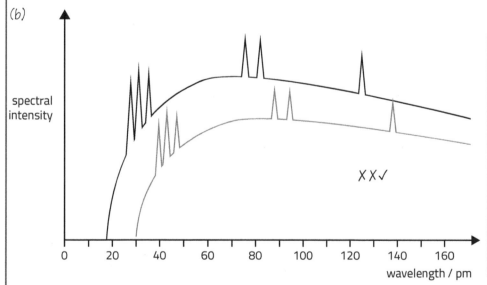

spectral intensity

wavelength / pm

✗ ✗ ✓

MARKER NOTE
Rhodri's answer looks ok but there are two important mistakes. First, all the peaks have moved to lower wavelengths. Second, the minimum wavelength should be 20 pm (increasing by 50% is increasing by a factor of 1.5, which means that the wavelength decreases by a factor of 1.5). He only gains the mark for the background spectrum.

1 mark

(c)   Fluoroscopy is expensive and useless because you don't want to put radioactive tracers inside the patient every year ✗. CTs give good images of the heart but has ionising radiation ✓. MRIs are expensive but don't have ionising radiation ✓.

Ultrasound is useless because the resolution isn't good enough ✗. PET scans are also low resolution but are far too expensive and not available in all hospitals. Overall, MRI would be best (unless the patient has a pace-maker).

MARKER NOTE
Rhodri has completely misunderstood fluoroscopy which would use iodine as a contrast medium but certainly has no radioactive tracer. He makes two acceptable points about CT scans (good image and ionising radiation). He also makes two acceptable points about MRI scans. He requires a little more to obtain the PET scan mark).

2 marks

Total	5 marks /11

Ffion's answers

(a) Photon energy $= \frac{hc}{\lambda}$
$= 6.63 \times 10^{-15}$ J ✓
$= 41.4$ keV ✓
So $V = 41.4$ kV ✓

MARKER NOTE

Ffion's answer is set out differently from the mark scheme. She first calculated the photon energy in J, converted this to eV which will then be the same (numerically) as the pd.

3 marks

(b)

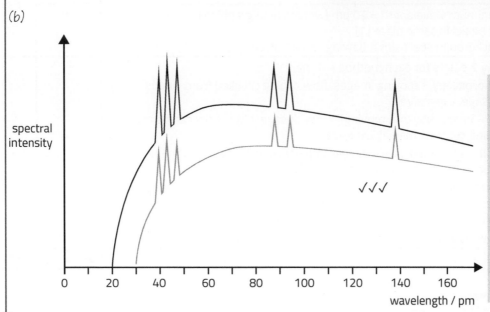

spectral intensity

✓✓✓

wavelength / pm

MARKER NOTE

Ffion's answer is awarded all 3 marks because she meets the criteria for all the individual marks. The only mark in doubt is the background radiation (this background continues underneath the peaks). The examiner has allowed it on this occasion because it is easier to draw the background spectrum first and add the peaks later.

3 marks

(c) The best and most cost-effective method would be to use non-ionising ultrasound B-scans (maybe even transoesophageal echocardiogram). These would provide cheap moving images and a Doppler ultrasound scan would give details of blood-flow too ✓. MRI is too expensive and all the other techniques involve ionising radiation ✓. Also, ultrasound is the only technique that will provide detailed blood-flow values and this is essential to check the operation of the replacement valve. Consultation over, my fee is $4000, my secretary accepts all major credit cards, thank you!! ☺ CT ✓, PET ✓

MARKER NOTE

Ffion has detailed knowledge about ultrasound scans and thoroughly deserves that mark. She has also identified that MRI is expensive and does not involve ionising radiation (this is implied but not stated). Because she has identified all the ionising and non-ionising radiations, she has made 1 good point about fluoroscopy, CT and PET but she goes further and states that none of these methods will give detailed blood-flow values. This is usually true of CT scans and definitely true of PET scans but not true of fluoroscopy where directions and speeds of flow can be calculated from the videos. Overall, she has 2 points about CT, PET, ultrasound and MRI but only 1 correct on fluoroscopy. Her final sentence, although amusing, will not be viewed favourably by examiners.

4 marks

Total   **10 marks / 11**

Option C: Physics of sport

Topic summary

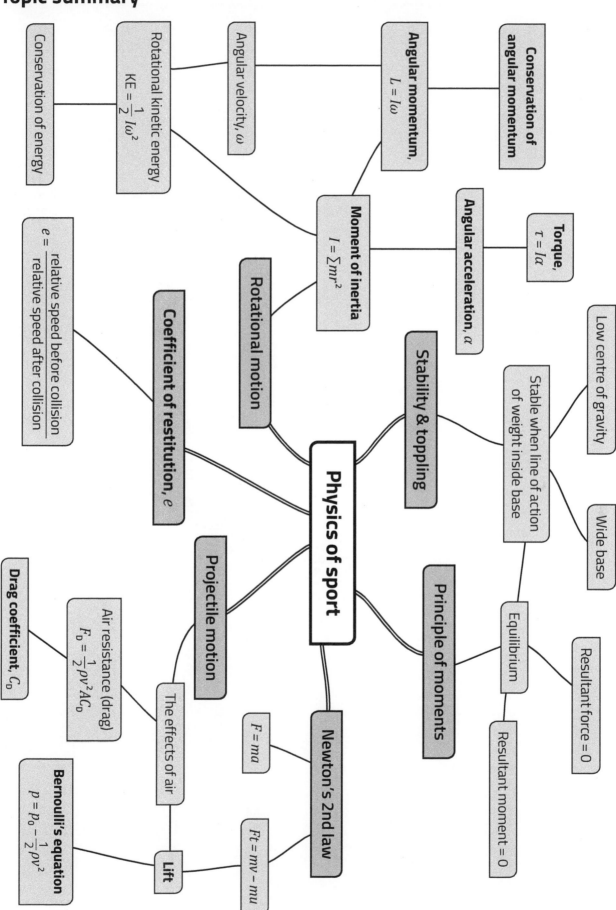

Conservation of energy

Rotational kinetic energy
$$KE = \frac{1}{2}I\omega^2$$

Angular velocity, ω

Angular momentum, $L = I\omega$

Conservation of angular momentum

Moment of inertia
$$I = \sum mr^2$$

Angular acceleration, α

Torque, $\tau = I\alpha$

$$e = \frac{\text{relative speed before collision}}{\text{relative speed after collision}}$$

Coefficient of restitution, e

Rotational motion

Stability & toppling

Stable when line of action of weight inside base

Low centre of gravity

Wide base

Physics of sport

Principle of moments

Equilibrium

Resultant force = 0

Resultant moment = 0

Projectile motion

Drag coefficient, C_D

Air resistance (drag)
$$F_D = \frac{1}{2}\rho v^2 A C_D$$

The effects of air

Newton's 2nd law

$F = ma$

$Ft = mv - mu$

Bernoulli's equation
$$p = p_0 - \frac{1}{2}\rho v^2$$

Lift

Q1 It is sometimes said that short, stocky players have a great advantage when playing football. Explain why this might be true, using stability. A simple diagram might help your answer. [3]

....

....

....

....

....

....

Q2 An athlete carries out a leg raise exercise as shown. The simplified diagram below shows the weight of the leg, the weight of the foot and the force (F) exerted on the leg by the relevant muscle. Calculate the force, F, required to hold the leg in equilibrium. [3]

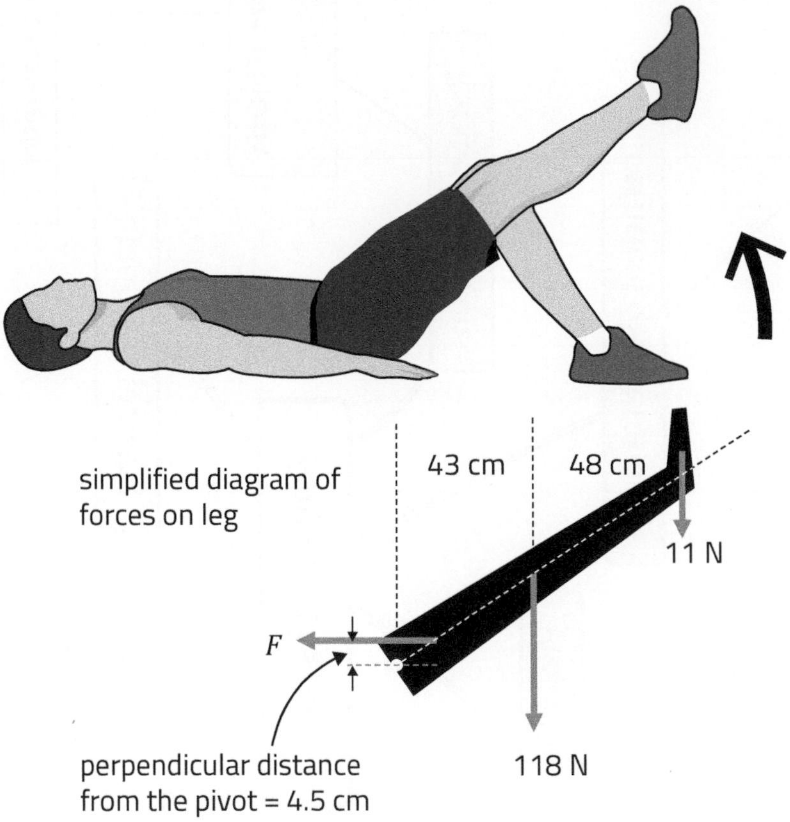

simplified diagram of forces on leg

43 cm   48 cm

11 N

F

perpendicular distance from the pivot = 4.5 cm

118 N

....

....

....

....

....

Q3 (a) A thin, hollow cylinder is rotated about its central axis.

(i) Explain why its kinetic energy (KE) is given by:

$$KE = \tfrac{1}{2}mr^2\omega^2$$

where r is the radius of the hollow cylinder, m its mass and ω its angular velocity. [3]

(ii) Show that, when the hollow cylinder rolls down an inclined plane from rest, half its kinetic energy will be linear KE and half will be rotational KE. [3]

(iii) When the hollow cylinder has rolled down an inclined slope a vertical distance of 0.30 m, its rotational KE is 0.45 J. Calculate the mass of the hollow cylinder. [3]

(b) The hollow cylinder and a snooker ball are both rolled down an inclined plane from the same height. Explain why the snooker ball arrives at the bottom of the inclined plane before the hollow cylinder (moment of inertia of a snooker ball = $\tfrac{2}{5}mr^2$). [3]

Q4 Serena Williams strikes a forehand topspin shot with extreme power. In the collision between racket and ball, the ball changes direction from 32 m s^{-1} due east to 48 m s^{-1} due west in a time of 6.8 ms. The ball also changes spin from 1200 revolutions per minute clockwise to 2550 revolutions per minute anticlockwise. The mass of the tennis ball is 58.2 g and its diameter is 66.9 mm.

(a) Calculate the linear and angular acceleration of the tennis ball. [5]

(b) Calculate the net torque and net force acting on the tennis ball during the collision.
[Moment of inertia of tennis ball = $\frac{2}{5}mr^2$] [4]

(c) Charles states that the rotational kinetic energy of the tennis ball is now greater than its linear KE. Determine whether, or not, he is correct. [3]

(d) Serena strikes the tennis ball at an angle of 6.5° above the horizontal from a height of 0.95 m above the ground. If the ball travels further than 31.2 m it will land beyond the court. Determine whether, or not, Serena has hit the ball too far.
[Ignore any effects of spin or air resistance for this part of the question.] [4]

(e) (i) The tennis ball has a drag coefficient of 0.60. Calculate the initial drag force acting on the tennis ball and comment whether, or not, drag can be ignored to a good approximation. [4]

(ii) The topspin on the tennis ball means that its lift is −2.0 N, i.e. there is a downward force of 2.0 N acting on the ball. Discuss how this force and the drag force might change your conclusion to part (d). [3]

Q5 When taking a shower after a match, a rugby player noticed that the shower curtain moved inwards when the shower was turned on. Explain this using the Bernoulli equation. [3]

Q6 The windsurfer shown is in equilibrium. Explain why the windsurfer does not fall backwards into the sea. [3]

Q7 Calculate the coefficient of restitution between snooker balls, using the data in the diagram: [3]

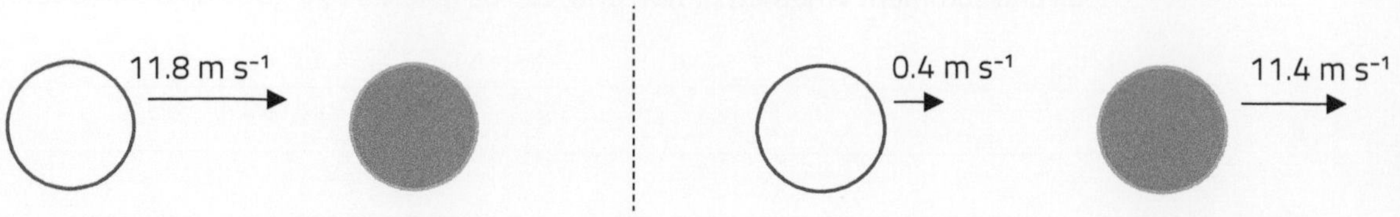

11.8 m s⁻¹ → 0.4 m s⁻¹ → 11.4 m s⁻¹ →

Q8 A trampolinist performs a triple somersault. She starts the jump with her body in the straight position but her rate of rotation increases when she goes into the tuck position. Explain why her rate of rotation increases when she goes into the tuck position. [3]

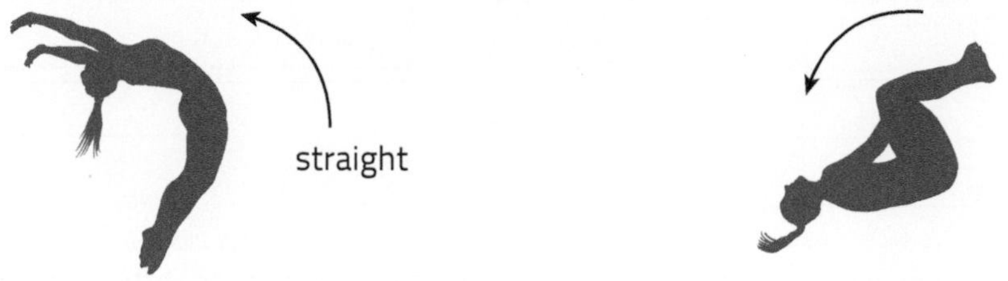

straight

tuck

Q9 (a) A tennis ball of mass 58 g is struck at a wall with a velocity of 63 m s⁻¹ due north. It rebounds off the concrete wall with a velocity of 49 m s⁻¹ due south. Calculate the mean force exerted by the wall on the ball given that the ball is in contact with the wall for 6.5 ms. [3]

(b) (i) Calculate the coefficient of restitution of the tennis ball and the concrete wall. [2]

(ii) To what fraction of its original height would you expect the tennis ball to rebound when dropped onto a concrete (hard court) floor? [2]

....

....

....

Q10 A golf club is in contact with a golf ball for 257 μs. During this time, a golf ball of mass 45.93 g and diameter 42.67 mm acquires a linear speed of 85 m s⁻¹ and a spin rate of 2700 revolutions per minute.

(a) Calculate the mean resultant force acting on the ball during the collision. [2]

....

....

....

(b) Calculate the mean angular acceleration of the golf ball. [3]

....

....

....

....

....

(c) Calculate the mean tangential force providing the ball's angular acceleration during the collision. [Moment of inertia of a solid sphere = $\frac{2}{5}mr^2$] [4]

....

....

....

....

....

....

(d) Draw the two components of the resultant force acting on the golf ball to scale in the directions indicated by the dotted lines. [2]

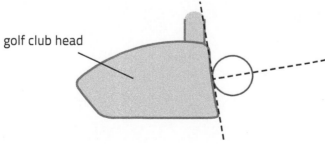

golf club head

(e) Calculate the initial ratio of rotational kinetic energy to linear kinetic energy for the golf ball. [3]

....

....

....

....

Q11 A football has a diameter of 22.0 cm. The ball is kicked by a goalie so that it moves from ground level at a speed of 18.1 m s^{-1} at an angle of 21.2° above the horizontal.

(a) Show that the ball should travel a horizontal distance of approximately 20 m before hitting the ground, if the effects of air resistance and spin are ignored. [3]

....

....

....

....

(b) The coefficient of drag of the football is 0.195. Calculate the initial drag force acting on the football. [ρ_{air} = 1.25 kg m^{-3}] [2]

....

....

....

(c) The ball is kicked from left to right with backspin. Explain why the actual distance travelled by the football is likely to be greater than your answer to part (a) [Hint: use the diagram]. [3]

direction of motion of football

....

....

....

....

....

(d) Draw labelled arrows to represent the three forces acting on the football (the ball has backspin so that the lift force is positive). [3]

direction of travel

(e) Evaluate the effects of spin and air resistance on the flight time of the ball, its mean horizontal speed and range. [6]

....

....

....

....

....

....

....

....

(f) The spin rate of the ball drops from 2700 revolutions per minute to approximately half this value when the ball lands. Explain whether or not this contradicts the principle of conservation of angular momentum. [2]

....

....

(g) The trajectory of the football in a vacuum is shown by the dotted line. Sketch the expected trajectory of the ball in air with backspin and air resistance (the expected range in air is 25 m). [3]

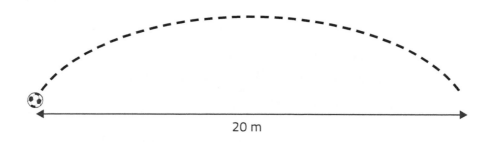

20 m

Question and mock answer analysis

Data relating to snooker balls can be seen in the table:

Snooker ball data		
Diameter	Mass	Drag coefficient
52.5 mm	165 g	0.48

A stationary snooker ball is struck with a cue tip. The tip is in contact with the ball for 1.15 ms during which time the ball accelerates to $13.6 \, m \, s^{-1}$.

cue tip

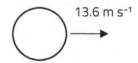

$13.6 \, m \, s^{-1}$

(a) Calculate the resultant force acting on the ball. [3]

(b) As the diagram shows, the snooker ball is not struck in its centre, so that a couple of 30.6 N m is exerted on the ball at the same time. Calculate the angular acceleration of the snooker ball. [3]

(c) The frictional force exerted by the cloth on the snooker ball is a constant and is given by:

$$\text{Frictional force} = 0.060 \times \text{weight of snooker ball}$$

(i) Calculate the speed at which friction and air resistance are equal for a snooker ball when the density of air is $1.25 \, kg \, m^{-3}$. [4]

(ii) Archibald claims that the effect of air resistance is greater than the effect of friction for the motion of a snooker ball. Discuss to what extent Archibald is correct. [3]

What the question is asking

Part (a) are mainly AO2 skills where Newton's 2nd law must be used to obtain the resultant force. Part (b) is very similar but is to do with rotational motion rather than linear motion. The moment of inertia of the snooker ball will have to be calculated too. Once again, (c)(i) are mainly AO2 skills and the equation for drag must be equated to the frictional force given in the question. Part (c)(ii) are AO3 skills. Friction and air resistance must be compared and then a logical conclusion should be presented

Mark scheme

Question part		Description	AOs			Total	Skills	
			1	2	3		M	P
(a)		Quoting or using Newton's 2nd law in either $F=ma$ or momentum form ($Ft = mv-mu$) [1]	1			3		
		Rearrangement or $\dfrac{0.165 \times 13.6}{0.00115}$ seen [1]		1			1	
		Correct answer = 1950 N [1]		1			1	
(b)		Substitution into moment of inertia equation i.e. $\frac{2}{5} \times 0.165 \times 0.02625^2$ (or 4.13×10^{-5} seen) [1]	1			3		
		$\alpha = \dfrac{\tau}{I}$, i.e. rearrangement (or implied) [1]		1			1	
		Correct answer = 673 000 rad s^{-2} [1]		1			1	
(c)	(i)	Use of drag equation i.e. $F = \frac{1}{2}\rho v^2 A C_D$ [1]	1			4		
		Use of weight equation i.e. $W = mg$ [1]		1				
		Equated the forces i.e. $0.06mg = \frac{1}{2}\rho v^2 A C_D$ [1]		1			1	
		Final answer = 12.2 m s^{-1} [1]		1			1	

	(ii)	Drag seems as large or comparable to friction (any comment comparing sizes) [1] Drag will decrease with decreasing speed (or converse) [1] Good final conclusion, e.g. drag more important at high speeds, overall fiction more important because drag drops (quickly), also accept both seem as important as each other (due to similar sizes) [1]				3	3		
Total			3	7	3	13	6	0

Rhodri's answers

(a) Force $= 0.165 \times 13.6$

$\qquad = 2.244\ N$ ✗

MARKER NOTE
Rhodri's answer only calculates the change in momentum of the snooker ball. There is no mark for this and he cannot be awarded any marks. Although he only needs to divide this answer by the time, none of the correct steps are present here.

0 marks

(b) Moment of inertia $= \frac{2}{3}mr^2$ ✗

$\qquad = 3.03 \times 10^{-4}$

$\alpha = \dfrac{30.6}{3.03 \times 10^{-4}}$ ✓

$\qquad = 101000\ rad\ s^{-2}$ ✗ no ecf

MARKER NOTE
Rhodri has used the incorrect equation for the moment of inertia (he has used the hollow sphere rather than the solid sphere equation). The method for the 2nd mark is correct.

1 mark

(c) (i) Drag $= \frac{1}{2} 1.25 v^2 (4\pi r^2) C_D$ ✗

Weight $= 9.81 \times 0.165 = 1.62$ ✓

$0.06 \times 1.62 = \frac{1}{2} 1.25 v^2 (4\pi \times 0.0525^2) C_D$ ✓

$\qquad v = 3.1\ m\ s^{-1}$ ✗ (no ecf)

MARKER NOTE
Rhodri has an excellent effort at this part but slips cost him two marks. He loses the 1st mark because he has used the surface area rather than the cross-sectional area and he has used the diameter instead of the radius. He obtains the 2nd mark for the weight equation and the 3rd for equating the forces. The final answer is wrong and no ecf can be awarded because the candidate has made a mistake in this section and not a previous section.

2 marks

(ii) If the forces are equal at such a low speed ✓(ecf) then it looks as though air resistance is more important than friction and Archie might just be correct. ✓ (ecf)

MARKER NOTE
Rhodri has a good effort at this part too and obtains 2 marks. He compares the forces (by saying they are equal at a low speed) and then comes to a very sensible conclusion (applying ecf for his low value in the previous part). He cannot gain the 2nd mark because there is nothing here relating to how the drag varies with speed.

2 marks

Total **5 marks /13**

Ffion's answers

(a) Force $= \dfrac{0.165 \times 13.6}{0.115}$ ✓✓

$= 2000 \, N$ ✓ BOD (rounded to 2sf)

> **MARKER NOTE**
> Ffion's answer is minimalist but the answer is correct and so is the method so she obtains full marks.
> **3 marks**

(b) $I = \frac{2}{5}mr^2$ ✓ $= 4.55 \times 10^{-5}$

$\alpha = \dfrac{30.6}{4.55 \times 10^{-5}}$ ✓ $= 672850 \, rad \, s^{-2}$ ✓

Must be wrong. Do a quick check.

$w = \alpha t = 774 \, rad \, s^{-1}$

So $f = 774 / 2\pi = 123 \, Hz$

Which seems quick but maybe ok.

> **MARKER NOTE**
> Ffion's answer is awarded all 3 marks because she meets the criteria for all the individual marks and the final answer is correct. She is not satisfied by the seemingly enormous angular acceleration and so does a quick check to see what the angular velocity would be (by multiplying the angular acceleration by the time). This was not required but was a very sensible thing to do when confronted by such a large number.
> **3 marks**

(c)(i) Drag $= \frac{1}{2} 1.25 v^2 (4\pi r^2) C_D$ ✓ BOD

$0.06 \times mg = = \frac{1}{2} 1.25 v^2 (4\pi \times 0.02625^2) C_D$ ✓✓

$v = 12.3 \, m \, s^{-1}$ ✓ BOD (incorrect rounding)

> **MARKER NOTE**
> Ffion's answer, once again, is succinct but the answer is correct and all the steps are shown (not that they are required when the answer is correct). She gains full marks once more.
> **4 marks**

(ii) If the forces are equal at such a low speed ✓ (ecf) then it seems to me that air resistance is more important than friction and Archie might just be correct. ✓ (ecf)

> **MARKER NOTE**
> Ffion has made an excellent effort at this part too and obtains 2 marks. She compares the two forces (by saying they are equal at a low speed) and then comes to a very sensible conclusion (applying ecf for his low value in the previous part). She cannot gain the 2nd mark because there is nothing here relating to how the drag varies with speed.
> **2 marks**

| Total | 12 marks /13 |

Option D: Energy and the environment

Topic summary

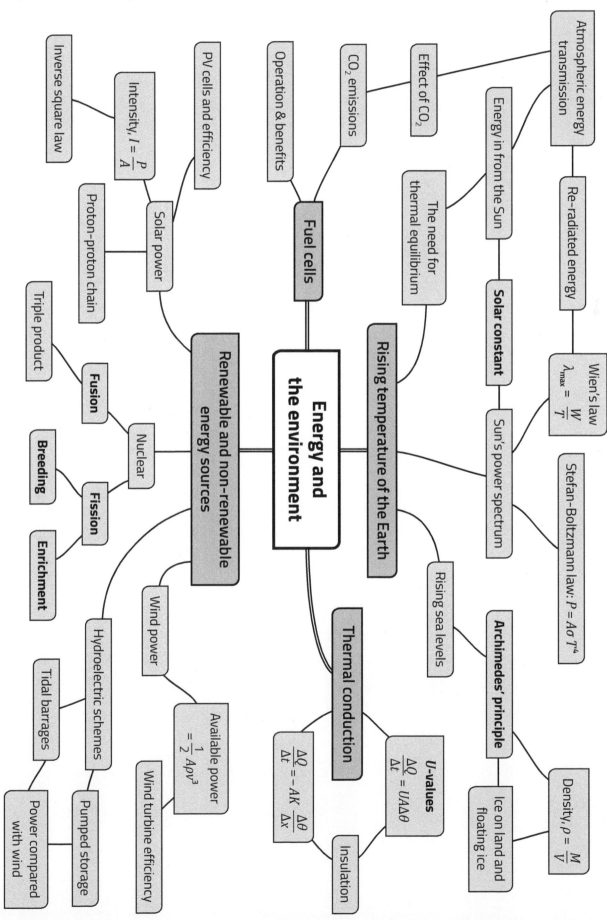

Q1 Asteroid 2010 TK7 has a diameter of 300 m. It orbits the Sun at the same distance as the Earth, 1.50×10^{11} m. The asteroid has no kind of atmosphere.
[Solar constant = 1361 W m^{-2}]

(a) The asteroid reflects 10% of the incident radiation back into space.

(i) Calculate the rate of absorption of solar energy by the asteroid. [2]

....

....

....

(ii) Assume that the asteroid is rotating and radiates as a black body. Show that the mean surface temperature of 2010 TK7 is approximately 270 K, explaining the significance of the assumptions. [4]

....

....

....

....

....

....

(iii) Calculate the peak wavelength of the radiation emitted from the surface of the asteroid, 2010 TK7. Identify the region of the electromagnetic spectrum in which this wavelength lies. [3]

....

....

....

....

(b) In spite of reflecting a bigger fraction of the incoming solar radiation, the mean surface temperature of the Earth is higher: approximately 288 K.

(i) Explain how the presence of greenhouse gases, such as carbon dioxide, in the atmosphere accounts for this. No calculations are needed. [4]

....

....

....

....

....

(ii) Explain briefly why increased concentrations of greenhouse gases are causing global temperatures to rise. [2]

....

....

....

Q2 (a) State the principle of Archimedes. [2]

....

....

....

(b) A block of ice has a volume of 100 cm³. It is lowered gently into a displacement can which is full of sea water.

ρ_{ice} = 920 kg m⁻³; $\rho_{sea\ water}$ = 1028 kg m⁻³

(i) Calculate the volume of seawater which is displaced by the ice. [3]

....

....

....

....

....

(ii) The ice in the can melts. Explain what happens to the water level in the can and the relevance of this for rising sea levels. [3]

....

....

....

....

....

(c) The area of ocean covered by sea ice and by glaciers has decreased because of global warming. This loss of ice accelerates global warming. Explain why this is the case. [2]

....

....

....

Q3 A common argument against the reliance on renewable sources for electrical energy generation is that they cannot be relied upon because they are intermittent. Discuss to what extent this is true and suggest how this drawback might be overcome. [4]

....

....

....

....

....

....

Component 3 Practice questions

Q4 (a) State what is meant by a *fuel cell*.   [2]

....

....

....

(b) Discuss the advantages in powering vehicles by fuel cells rather than petrol or diesel engines.   [3]

....

....

....

....

....

Q5 (a) State what is meant by uranium *enrichment* and explain why it is necessary.   [3]

....

....

....

....

....

....

(b) Fissile nuclides can be bred (i.e. produced) when a non-fissile nuclide absorbs a neutron from a fission reactor. This produces a radioactive nuclide which undergoes multiple β^- decays to give the desired fissile nuclide.

The fissile $^{233}_{92}U$ can be bred from an isotope of thorium, Th, for which Z = 90.

(i) Write the reactions for breeding $^{233}_{92}U$.   [2]

(ii) Describe how a fissile nuclide of plutonium is produced.   [3]

....

....

....

....

....

....

Q6 (a) Air, of density, ρ, travels at a speed, v, along a pipe of cross-sectional area A.
Show that the kinetic energy of the air which passes any point in the pipe per second is given by
$\frac{1}{2}A\rho v^3$. A space is included for a diagram, if needed. [3]

(b) A wind turbine with blade length 80 m operates in a steady wind of speed 12 m s^{-1}. It generates
7.9 MW of electrical power of. Calculate its efficiency. [ρ_{air} = 1.25 kg m^{-3}] [3]

Q7 The thermal conduction equation can be written:

$$\frac{\Delta Q}{\Delta t} = -AK\frac{\Delta\theta}{\Delta x}.$$

(a) Explain what the following components of the equation refer to and give their units: [3]

$\frac{\Delta Q}{\Delta t}$ ....

....

$\frac{\Delta\theta}{\Delta x}$ ....

....

(b) Hence, show that the unit of thermal conductivity, K, is W m^{-1} K^{-1}. [2]

(c) State the significance of the minus sign (–) in the equation. [1]

(d) Calculate the heat transfer per minute through a 0.50 m^2 panel of pine of thickness 12.0 mm if there is
a temperature difference of 25 °C across the panel. [3]
[K_{pine} = 0.14 W m^{-1} K^{-1}]

Q8 A domestic photovoltaic (PV) system consists of a set of individual PV panels. The manufacturer produces the following characteristic graphs for the 2.0 m² individual panels illuminated at 90° by solar radiation at different intensities.

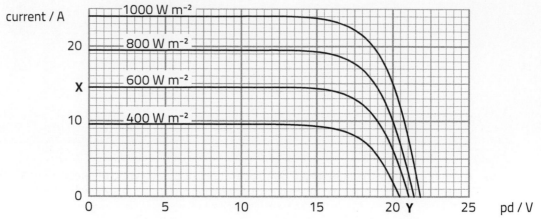

(a) The solar constant is 1361 W m⁻². Give a reason why the manufacturer's graphs only go up to 1000 W m⁻². [1]

(b) Use the data to confirm that the output current at a particular pd (e.g. 15 V) is proportional to the radiation intensity. [2]

(c) The point **X** represents the output for 600 W m⁻² when the terminals on the PV panel are short circuited. At **Y** the terminals are open circuited. Explain why the power output is zero at both X and Y. [2]

(d) The manufacturer claims that the panel can give out at least 220 W when in light of intensity 600 W m⁻². Evaluate this claim and determine the output pd and current which deliver the maximum power. [4]

Q9   In the small-scale hydroelectric scheme shown, 2.75 m³ of water flows through the turbine each second; the diameter of the outflow pipe is 1.0 m².

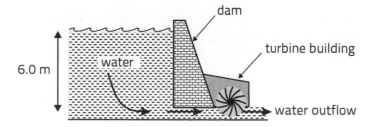

(a)   Show that:

(i)   the rate of gain of kinetic energy by the water is approximately 17 kW,   [3]

(ii)   the rate of loss of potential energy by the water is approximately 160 kW.   [2]

(b)   Assuming that the turbine/generator combination converts 80% of the available energy to electrical energy, calculate the output power of the generator.   [3]

(c)   Colin claims that if the turbine is adjusted to allow a 10% greater flow rate of water, the power output would be 10% more but overall efficiency of the hydroelectric generation would be 10% less. Evaluate these claims.   [5]

Q10 The first two stages of the proton-proton chain are:

$$^1_1H + ^1_1H \rightarrow ^2_1H + ^0_1e + ^0_0\nu_e \quad \text{and} \quad ^2_1H + ^1_1H \rightarrow ^3_2He + \gamma$$

The mean lifetime of protons in the core of the Sun is over 1000 million years, whereas a deuteron, ^2_1H, reacts to give ^3_2He within about 1 second. Briefly explain this difference by considering the interactions involved in the two stages.

[2]

....

....

....

Q11 The exterior of a flat consists of an insulated wall with a U-value of $0.18 \text{ W m}^{-2} \text{ K}^{-1}$ and two identical double-glazed windows with U values of $1.5 \text{ W m}^{-2} \text{ K}^{-1}$.

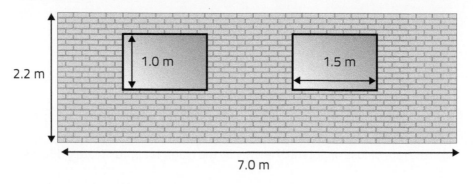

(a) Explain what is meant by a U-value of $0.18 \text{ W m}^{-2} \text{ K}^{-1}$.

[2]

....

....

....

(b) The flat owner is considering replacing the double-glazed units with triple glazing which has a U-value of $0.8 \text{ W m}^{-2} \text{ K}^{-1}$. Assuming that the flat only loses heat through the wall and windows, calculate the percentage reduction in heating bill which the owner should expect.

[3]

....

....

....

....

....

(c) Suggest why triple-glazed windows are much more common in Norway than in Wales and England. [2]

....

....

....

Q12 The wall of a building consists of two 10 cm brick layers with a 10 cm layer of mineral wool insulation between. On one day, the temperature difference across the interior brick layer is 0.35 °C.
$K_{brick} = 0.62$ W m^{-1} K^{-1}; $K_{insulation} = 0.039$ W m^{-1} K^{-1}

(a) Calculate the rate of heat loss from an 8.0 m^2 section of wall. [2]

(b) Show that the temperature difference across the whole wall is approximately 6 °C. [2]

(c) On a windless day, the temperature of the air inside the building is 22 °C and that of the outside air is 8 °C.

(i) Determine the U-value of the wall. [3]

(ii) With the aid of a diagram but without any calculations, explain why the temperature difference between the inside and outside air temperatures is much greater than the 6 °C given in part (b). [3]

(iii) Explain briefly why you would expect the rate of heat loss on a windy day to be greater than predicted by the U-value. [2]

Q13 The graph shows the variation of intensity with wavelength of the Sun's radiation. [Note the logarithmic scale.]

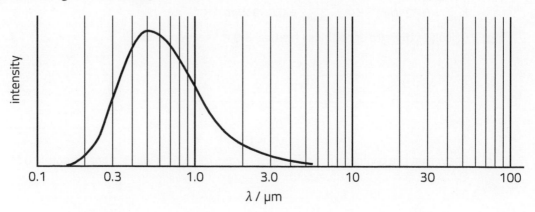

(a) Show that the temperature of the surface of the Sun is approximately 6000 K. [2]

....

....

....

(b) The mean temperature of the Earth's surface is about 290 K. Sketch the emission spectrum of the Earth on the same axes. [The height of the peak is not important.] [2]

[Space for calculations]

(c) Radiation between 0.4 and 0.9 μm passes through the atmosphere with little absorption. Between 3 and 10 μm, a significant fraction of radiation is absorbed by carbon dioxide, water vapour and methane molecules. At longer wavelengths virtually all is absorbed.

Explain the significance of this information for the temperature of the surface of the Earth. [5]

....

....

....

....

....

....

....

....

Q14 It is planned to place a thermal store in a cupboard which has a mean temperature of 20°C. The store is to be used for providing hot water for domestic use. The thermal store consists of a stainless steel cylinder of diameter 0.60 m and thickness 0.3 cm, containing 400 litres of water. The cylinder is insulated by a 2.5 cm thick polyurethane (PU) foam jacket and stands on an insulating base.

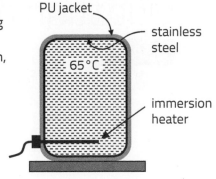

A domestic PV system is used to maintain the temperature of the water at 65°C, via an immersion heater.

$K_{PU} = 0.025$ W m^{-1} K^{-1}; $K_{steel} = 45$ W m^{-1} K^{-1}; 1 litre = 10^{-3} m^3

(a) Calculate the surface area of the PU foam ignoring the base. [3]

(b) In calculating the rate of heat loss from the thermal store, Damian commented that you could ignore the effect of the steel. Justify this comment. [2]

(c) (i) Use this information to show the water will need to be heated at a rate of about 130 W by the PV system to maintain its temperature. [2]

(ii) Estimate the temperature difference across the stainless steel and consider again whether ignoring the stainless steel in part (i) was justified. [2]

(d) When installed, it is noted that the actual heating rate when heat is not being extracted from the store for domestic use is only 70 W. Discuss what was ignored in the calculation in (c)(i). [2]

Question and mock answer analysis

Q&A 1

(a) The first stage in the fusion reactions taking place in the Sun is:

$$^1_1H + {}^1_1H \rightarrow {}^2_1H + {}^0_1e + {}^0_0v_e$$

with the release of 2 MeV.

Suggest **two** reasons why this reaction is not considered appropriate for use in a practical nuclear fusion reactor. [2]

(b) An important concept in nuclear fusion is that of *confinement time*, τ_E, defined by:

$$\tau_E = \frac{W}{P_{loss}}$$

where W is the energy density (i.e. the energy per unit volume) of the plasma and P_{loss} is the rate of energy loss per unit volume.

Show that, with this definition, τ_E has the unit of time. [2]

(c) In order to produce useful power, the *triple product* must reach a critical value.

   (i) The unit of the triple product can be written $K\,s\,m^{-3}$. Identify the three quantities in the triple product and hence justify this unit. [3]

   (ii) The critical value of the triple product depends on the temperature. A school physics book gives the following graph of the variation of the critical value of the triple product with temperature for a deuterium-tritium reactor.

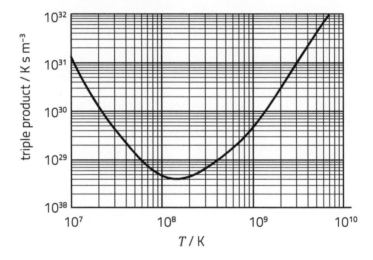

   (I) Give the minimum value of the triple product needed to sustain a fusion reaction and the temperature at which this occurs. [2]

   (II) In an experimental reactor, a confinement time of 2.0 s is achieved when working at the temperature in part (I). Calculate the density of deuterium and tritium ions necessary to achieve sustained fusion. [2]

What the question is asking

There is very little that examiners can ask about nuclear fusion in this option but part (a) is synoptic in nature and is AO3. You are expected to realise that, as this a weak reaction, it is very unlikely to occur and it also gives out very little energy. The triple product is explicitly mentioned in the option specification. Parts (b) and (c) of the question combine knowledge of the triple product with that of SI units and data analysis. Part (b) gives a definition of confinement time and requires its use in deriving a unit – a standard AO2 question. Part (c) (i) extends this to asking for a recall of the definition of triple product (AO1) and justifying its unit (AO2). The data analysis on the triple product in (c)(ii) is quite tricky not only because it introduces the idea that the triple product (which includes temperature) varies with temperature but also because the data is given using a graph with logarithmic scales, which is one of the mathematical requirements.

Mark scheme

Question part			Description	AOs			Total	Skills	
				1	2	3		M	P
(a)			It is a weak interaction so unlikely / fewer interactions (per second)[1] Its energy release is low (cf. ^{2}H + ^{3}H)[1]			2	2		
(b)			$[W]$ = J m^{-3} [or kg m^{-1} s^{-2}] **and** $[P_{loss}]$ = W m^{-3} [or kg m^{-1} s^{-3}] [1] Division seen → s [1] Note: omission of m^{-3} → 1 mark max.		2		2	1	
(c)	(i)		Containment time, particle/ion density, temperature [1] $nT\tau_E$ seen [or in words] [1] m^{-3}, K seen [1]	1 1	1		3	1	
	(ii)	I	4×10^{28} [K s m^{-3}] 1.3×10^8 [K] [allow 1.2 − 1.4]	2			2		
		II	4×10^{28} (ecf) = $n \times 1.3 \times 10^8$ (ecf) × 2.0 [1] 1.5×10^{20} [m^{-3}]	1	1		2	1	
Total				5	4	2	11	3	0

Rhodri's answers

(a) The protons would just bounce off each other and not react because they are positively charged X [not enough]

2 MeV = 3.2 × 10^{-13} J X (?)

MARKER NOTE
Rhodri needed to link this to the weak interaction to gain the 1st mark. ^{2}H and ^{3}H ions are also positively charged! What is the significance of 3.2 × 10^{-13} J?

0 marks

(b) Energy (W) has unit J
Power (P) has unit W X

$W = J s^{-1}$ ∴ $\dfrac{J}{J s^{-1}}$ = s ✓ ecf

MARKER NOTE
Rhodri has not noticed that W and P_{loss} are per unit volume, so loses the first mark. He is allowed the second mark ecf.

1 mark

(c) (i) Apart from the containment time, there is the number of nuclei per cubic metre and the temperature. ✓
The units are s, m^{-3} and K ✓
So we get K s m^{-3} X [not enough]

MARKER NOTE
Rhodri has identified the three variables for the first mark. He has given the units in the same order as the variables so the examiner assumes that the m^{-3} and K are correctly ascribed. He needs to say that the quantities are multiplied together for the middle mark.

2 marks

(ii) (I) Triple product = 3 × 10^{28} X

Temperature = 0.5 × 10^8 X

MARKER NOTE
Rhodri has not interpreted the scale correctly, e.g. the line after 10^8 is 2 × 10^8 so the temperature at minimum is between 1 and 2 × 10^8.

0 marks

(II) Density = $\dfrac{3 \times 10^{28}}{0.5 \times 10^8 \times 2.0}$ ✓ (ecf)

= 3 × 10^{20} m^{-3} ✓

So 1.5 × 10^{20} each of deuterium and tritium.

MARKER NOTE
The examiner has allowed ecf on the answer to part (b)(ii)I. Dividing the density into deuterium and tritium wasn't required but certainly isn't penalised!

2 marks

Total   **5 marks /11**

Component 3 Practice questions

Ffion's answers

(a) The neutrino shows that this is a weak interaction. So, if when lots of protons collide very few will react like this and change into ^2_1H so very little energy will be produced. ✓

Also a lot of the energy will be taken away by the neutrinos – this will be lost to the reactor. ✓

MARKER NOTE
This is an excellently developed answer on the first marking point only. Unfortunately the mark scheme does not allow two marks to be given for this. However, Ffion's second point which is not on the mark scheme is very insightful and would have been awarded after consultation with a senior examiner.

2 marks

(b) Energy per unit volume $= kg\ m^2\ s^{-2} \times m^{-3}$
$$= kg\ m^{-1}\ s^{-2}$$

So power loss per unit volume $= kg\ m^{-1}\ s^{-3}$ ✓

So $\left[\dfrac{W}{P_{loss}}\right] = \dfrac{kg\ m^{-1}\ s^{-2}}{kg\ m^{-1}\ s^{-3}}$ ✓ $= s$

MARKER NOTE
The answer is fine, though Ffion didn't need to express [W] and [P_{loss}] in base SI units.

2 marks

(c)(i) The triple product:
Containment time, $\tau_E - s$
number density, $n - m^{-3}$ [bod]
temperature, $T - K$ ✓✓
So $[T\tau_E n]$ ✓ $= K\ s\ m^{-3}$

MARKER NOTE
The only weakness in Ffion's answer is that she doesn't say what the 'number' is. Either reacting nuclei or ions is expected – hence the bod.

3 marks

(ii)(I) Minimum value $= 4 \times 10^{28}\ K\ s\ m^{-3}$ ✓
Temperature $= 1.5 \times 10^8\ K$ ✗

MARKER NOTE
Ffion has basically understood the way that a logarithmic scale works and got the easy first mark. She has not followed through with applying the non-linearity of the scale in the more difficult second mark.

1 mark

(II) $4 \times 10^{28} = 1.5 \times 10^8 \times 2.0\ n$ ✓ (ecf)
$\therefore n = 1.3 \times 10^{20}\ m^{-3}$ ✓

MARKER NOTE
The examiner has allowed ecf on the incorrect temperature, so full marks.

2 marks

Total　　　　　　　　　　**10 marks /11**

Practice papers

A LEVEL PHYSICS
COMPONENT 3 PRACTICE PAPER

2 hours 15 minutes

		For Examiner's use only	
	Question	Maximum Mark	Mark Awarded
Section A	1.	7	
	2.	8	
	3.	6	
	4.	23	
	5.	9	
	6.	19	
	7.	16	
	8.	12	
Section B	Option	20	
	Total	120	

Notes

This paper is in two sections, **A** and **B**.

Section **A**: 100 marks. Answer **all** questions. You are advised to spend about 1 hour 50 minutes on this section.

Section **B**: 20 marks; Options. Answer **one option only**. You are advised to spend about 25 minutes on this section.

In an Eduqas paper, the following information will be given on the front of the paper:

1. **Additional materials**
   You will be told that you will require a calculator and a Data Booklet. Sometimes you will be told that you need a ruler and/or an angle measurer / protractor.

2. **Answering the examination**
   You will be told to use a blue or black ball-point (but graphs are best drawn using a pencil).
   You will be told to answer all the questions in the spaces provided on the question paper.

3. **Further information**
   Each question part shows, using square brackets, the total marks available. One question will assess the quality of extended response [QER]. This question will be identified on the front page. In this practice paper the QER question is question **3**.

SECTION A

Answer all questions.

1. (a) Explain the difference between a longitudinal wave and a transverse wave. [2]

....

....

....

   (b) Hence, explain why only a transverse wave can be polarised. [1]

....

....

   (c) Light from a distant lamp is reflected off glass. A physics student analyses the reflected light using a rotating polaroid and a light intensity app on her mobile phone. She also analyses light directly from the lamp.

Explain what conclusions the student should draw regarding the percentage of polarised light in the direct and the reflected light in this experiment. You should include values in your answer. [4]

....

....

....

....

....

....

....

2.   A specified practical for GCSE Physics involves analysing moving water waves in a ripple tank. Students are required to estimate the wavelength of moving waves using a ruler and to count the number of waves passing a point in 10 s. Here are Jemima's results:

Length of 5 waves / cm				Wavelength / cm
Reading 1	Reading 2	Reading 3	Mean	
8	7	7		
Number of waves passing in 10 s				Frequency / Hz
Reading 1	Reading 2	Reading 3	Mean	
33	36	38		

(a) Complete the table.   [2]

(b) Calculate the speed of the water waves.   [2]

....

....

....

(c) Jemima states, 'I can get results that are 10 times more precise by using my mobile phone to take a photograph of the waves and also to take a video of the waves.' Use data from the table to justify whether, or not, she is right in her claim.   [4]

....

....

....

....

....

....

....

....

3. Explain the difference between stationary waves and progressive waves, paying particular attention to energy, amplitude and phase. In your answer you should discuss one specific example of a stationary wave and one specific example of a progressive wave. [6 QER]

....

....

....

....

....

....

....

....

....

....

....

....

....

....

....

4. (a) Explain how two in-phase wave sources can produce an interference pattern. [3]

....

....

....

....

....

(b) An experiment is carried out to determine the wavelength of laser light using Young's double slit experiment. The following set-up is used:

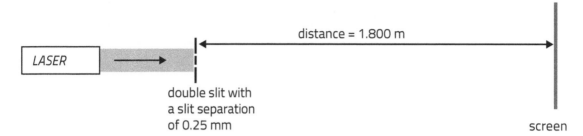

Samir carries out the experiment by marking the centre of each bright fringe. He then measures the distance from the central fringe to each fifth fringe, recording the data in the table.

Fringe number	Fringe distance (from centre of central fringe) / cm
5	2.1 ± 0.1
10	3.9 ± 0.1
15	5.6 ± 0.1
20	7.3 ± 0.2
25	9.2 ± 0.1

(i) Use the $n = 25$ fringe data to show that the laser wavelength is in the visible part of the electromagnetic spectrum. [3]

....

....

....

....

....

(ii) Samir states that the location of the 20th fringe was difficult to obtain precisely because every fourth fringe was very faint, i.e. the 4th, 8th, 12th, 16th, 20th and 24th fringes were faint. Suggest a reason why every 4th fringe has a low intensity. [2]

....

....

....

(iii) Plot a graph of fringe distance against fringe number. Include error bars for the fringe distance. [5]

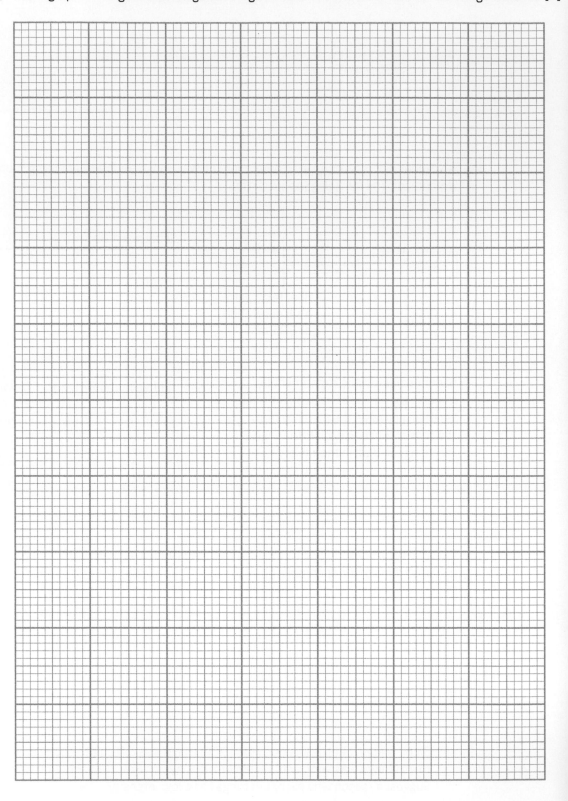

(iv) Explain why the graph suggests that there is a systematic error in the data and suggest a reason for this systematic error. [2]

....

....

....

....

(v)   Use the graph to calculate the wavelength of the laser used in the experiment along with its absolute uncertainty. You may assume that the slit separation and the distance from the slit to the screen have negligible uncertainties.   [5]

(vi) Explain why the wavelength of the laser can be obtained more precisely using a diffraction grating rather than a double slit.   [3]

5. (a) Explain the difference between spontaneous emission and stimulated emission. [3]

....

....

....

....

(b) Explain, briefly, the importance of a population inversion in a laser system. [3]

....

....

....

....

(c) Calculate the emission wavelength of the laser whose principal energy levels are shown. [3]

35.0 eV ——————————— Lifetime = 3 ns

31.8 eV ——————————— Lifetime = 0.8 ms

29.6 eV ——————————— Lifetime = 5 ns

0 eV ——————————— Stable

....

....

....

....

....

6. (a) (i) Fluorine-18 ($^{18}_{9}F$) undergoes β^+ decays to oxygen (O). Complete the nuclear equation for the decay process. [3]

$$^{18}_{9}F \longrightarrow$$

(ii) Explain how baryon number, charge and lepton number are conserved in this reaction. [3]

....

....

....

....

....

(iii) State with reasons which force is responsible for this decay process. [3]

....

....

....

....

....

(b) (i) The half-life of $^{18}_{9}F$ is 109.8 minutes. Calculate the mass of fluorine-18 required to produce an activity of 370 MBq. [4]

....

....

....

....

....

(ii) Calculate the time taken for the activity to decrease from 370 MBq to 20 MBq. [3]

....

....

....

....

....

(c) Each emitted β^+ particle comes to rest and undergoes mutual annihilation with an electron in the patient's body, with the emission of two γ photons.

Explain why these photons are emitted in opposite directions and have equal energy. [3]

....

....

....

....

7.   **'Soft errors' in computer chips**

Boron-containing glass is used as an insulator between conducting layers of computer chips.

'Thermal' neutrons are low energy neutrons (~ 1/40 eV) produced by the collision of cosmic rays (high energy protons from space) with atmospheric atoms. They can cause damage to information stored on computer chips if they are absorbed by ^{10}B, which comprises 20% of naturally occurring boron. This is called a soft error.

The reaction is $^{10}_{5}B + ^{1}_{0}n \rightarrow ^{11}_{5}B^*$

The $^{11}_{5}B^*$ atom decays almost straightaway, with the emission of an alpha particle ($^{4}_{2}He$), to an isotope of lithium (Li).

(a) Complete the decay equation for $^{11}_{5}B^*$. [2]

$$^{11}_{5}B^* \rightarrow$$

(b) (i) Use the following mass data to calculate the energy released in the decay. [3]

Mass/u: $^{10}_{5}B = 10.013\ 54$; $^{1}_{0}n = 1.008\ 67$; lithium isotope = 7.016 00; $\alpha = 4.002\ 60$

(ii) Explain why it is not necessary to take account of the kinetic energy of the thermal neutron in the above calculation. [2]

(c) The alpha particle is emitted with approximately two thirds of the total energy of the decay. Without calculation, explain why it has more kinetic energy than the lithium atom. [3]

(d) The * in the symbol $^{11}_{5}B^*$ indicates that the ^{11}B is formed in a high energy state. Normal $^{11}_{5}B$, which is 80% of natural boron, has a mass of 11.009 31 u. Explain why it is does not undergo alpha decay. [2]

(e) The ionisation caused by the α particles can change the voltage levels in the computer chips, effectively introducing errors in the information stored.

(i) If ^{11}B absorbs a neutron it then undergoes β$^-$ decay to form ^{12}C. Suggest why this is less of a problem for computers. [2]

....

....

....

(ii) ^{11}B comprises 80% of natural boron. Up to now, computer manufacturers have not used glass with the 20% of ^{10}B removed. Discuss why they have not done this. [2]

....

....

....

8. A conducting rod, **XY**, of resistance 0.50 Ω and mass 0.20 kg, slides at a constant speed of 8.0 m s⁻¹ on a pair of frictionless conducting tracks, **AB** and **CD**, as shown. The tracks have negligible resistance. The plane of the tracks and the direction of motion of the conductor are both at right angles to the magnetic field.

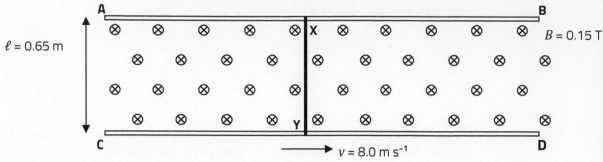

(a) (i) Calculate the emf induced in **XY**.   [1]

....

....

   (ii) Explain which rail, **AB** or **CD**, acquires a positive charge.   [2]

....

....

....

   (iii) Explain why there is no resistance to the motion due to the induced emf.   [2]

....

....

....

(b) A conductor of negligible resistance is connected between **A** and **C**.

   (i) Calculate the rate of dissipation of energy in the circuit.   [2]

....

....

....

   (ii) A force parallel to the rails is now applied to the rod via a string attached to its midpoint. The force keeps the rod moving at 8.0 m s⁻¹. Calculate the tension in the string.   [2]

....

....

....

   (iii) The string is now removed and the rails are inclined at a small angle, θ (< 5°). The rod **XY** is observed to move at the same constant rate of 8.0 m s⁻¹. Calculate θ.
        Hint: $\cos \theta = 1.00$ (to 3 sf).   [3]

....

....

....

....

SECTION B: OPTIONAL TOPICS

Option A – **Alternating currents** ☐

Option B – **Medical physics** ☐

Option C – **The physics of sports** ☐

Option D – **Energy and the environment** ☐

Answer the question on one topic only.

Place a tick (✓) in one of the boxes above, to show which topic you are answering.

You are advised to spend about 25 minutes on this section.

Option A – Alternating currents

9. (a) A rectangular 5.0 cm × 3.0 cm plane coil consists of 50 turns of wire. A teacher sets up the coil as a model generator, with the coil at right angles to a uniform magnetic field of 0.20 T. She rotates the coil with a frequency of 2.5 Hz.

Sketch a graph of the expected output voltage, V_{out}, against time over two complete revolutions, with $t = 0$ when the coil is at 90° to the field. Include values on the axes. [3]

Space for calculations

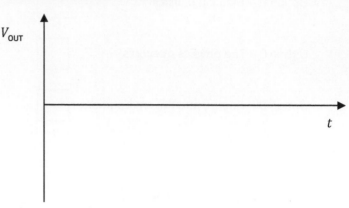

(b) A series RCL circuit is set up. The frequency of the supply is f_1. The impedances of the three components at this frequency are shown in the diagram:

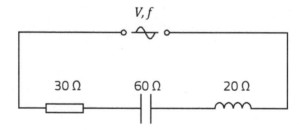

The current in the circuit is 0.40 A.

(i) Explain why current would be the same if the frequency were increased to $3f_1$, keeping the rms voltage constant. [2]

....

....

....

....

(ii) (I) Sketch the resonance curve for the circuit, paying close attention to the resonance frequency.

[3]

Space for calculations

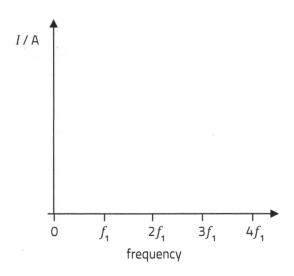

(II) The 30 Ω resistor is now replaced by a 50 Ω resistor.

Without further calculations, explain how the *Q* factor of the new circuit compares with that of the first circuit.

[2]

....

....

....

(c) Paula used a double beam CRO to measure the value of an inductor in this AC circuit.

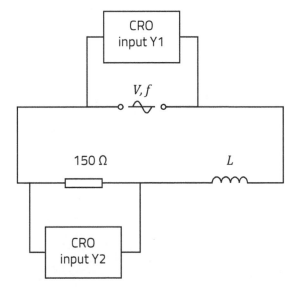

She connected the Y1 inputs of the CRO across the terminals of the signal generator (the power supply) and the Y2 inputs across the 150W resistor.

She set the Y gains and the time base as follows:

Y1: 5 V div⁻¹
Y2: 2 V div⁻¹
Time base : 50 µs div⁻¹

With these settings she obtained the following traces on the CRO:

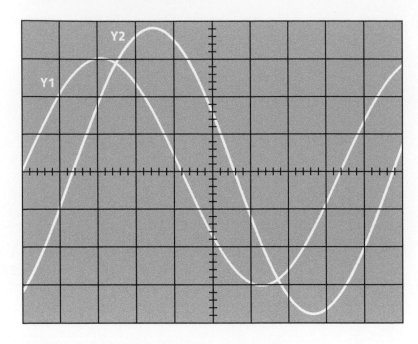

(i) Determine the peak voltage and frequency of the supply. [2]

....

....

....

....

(ii) Show that the peak current in the circuit is approximately 50 mA. [2]

....

....

....

....

(iii) Determine the impedance of the circuit and hence the inductance, L, of the inductor. [3]

....

....

....

....

....

(iv) Evaluate whether the phase difference between the supply voltage and the current is consistent with values of the inductance and resistance. [3]

....

....

....

....

Option B – Medical physics

10. (a) The diagram shows an X-ray spectrum produced by a metal target.

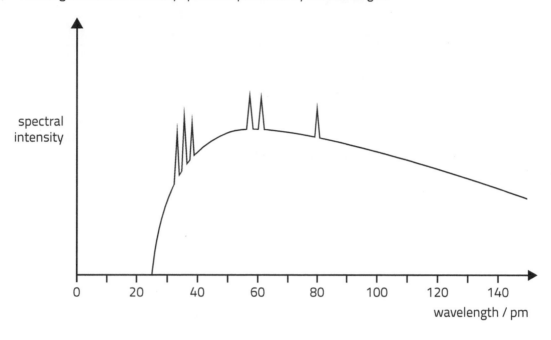

spectral intensity

wavelength / pm

(i) Explain how the continuous spectrum or 'bremsstrahlung' arises. [1]

(ii) The tube current is 55 mA and 37 W of X-ray output is produced. Calculate the efficiency of the X-ray tube. [4]

(iii) **On the diagram**, sketch the X-ray spectrum for the same metal target when the tube voltage is halved. [2]

(b) An ultrasound technician uses ultrasound of frequency 6.4 MHz to measure the speed of blood flow in a 6-month-old foetus. The ultrasound enters the aorta almost exactly parallel to the blood flow and a frequency shift of 6.6 kHz is detected.
Calculate the speed of blood flow in the foetus's aorta.
[The speed of ultrasound in blood is 1570 m s⁻¹.] [3]

(c) (i) Explain the difference between *absorbed dose* of radiation for an organ and the *equivalent dose* for an organ. [2]

....

....

....

(ii) The total effective dose due to radon gas is the effective dose to the lungs only. This is because it is inhaled. Radon gas emits alpha particles and the annual absorbed dose received by the lungs of an unfortunate inhabitant of an old basement flat in Cornwall is 0.140 Gy. Use the table to calculate the effective dose to the inhabitant of the basement flat. [3]

Radiation type	Radiation weighting factor	Tissue	Tissue weighting factor
Alpha	20	Gonads	0.25
Beta	1	Lungs	0.12
Gamma	1	Red bone marrow	0.12
Protons	5	Thyroid	0.03
Neutrons <10 kV	5	The rest of body	0.48

....

....

....

....

....

(d) A patient is suspected of having a spinal-cord injury. Evaluate the effectiveness of the following imaging techniques in diagnosing the injury:

Ultrasound   X-ray   CT scan   MRI   PET scan   [5]

....

....

....

....

....

....

....

....

....

....

Option C – The physics of sports

11. (a) Use the diagram of a rectangular block standing on its end to explain why a Formula 1 (F1) car is much more stable against toppling than a double-decker bus. [3]

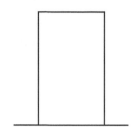

....

....

....

....

....

....

....

(b) (i) A good bowler can make a cricket ball swing as it travels through the air. According to a website, the explanation for this is to do with the raised seam, the rough side and the polished side of the ball.

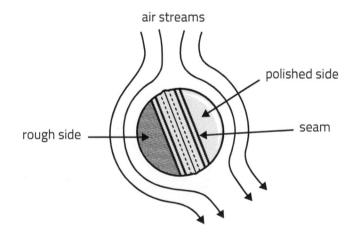

air streams

polished side

rough side

seam

The theory is that the air stream is held longer on the rough side than on the polished side so it changes direction as shown. The seam assists the separation of the air flow and the ball on the polished side.

Explain, in terms of Newton's laws of motion, how this will affect the change in the direction of motion of the ball. [4]

....

....

....

....

....

....

(ii) A batter optimistically despatches a cricket ball from close to ground level at an angle 25° to the horizontal, and a speed of 32.0 m s⁻¹ towards the 75 m distant boundary rope at the Headingley cricket ground.

A fielder is standing in the correct position just inside the boundary rope and will catch any ball lower than 2.40 m. Evaluate whether the batter scores a boundary, is caught out or neither. Show your working clearly. [5]

(c) The diagram shows a hammer thrower about to release the hammer. In order to score a personal best, he needs to release the hammer at a speed of 28 m s⁻¹.

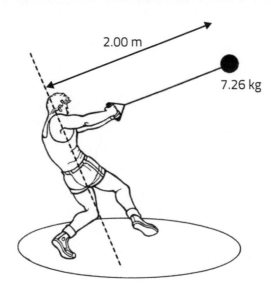

2.00 m

7.26 kg

(i) Using information from the diagram, calculate the moment of inertial of the hammer about the central axis of the hammer thrower. [2]

(ii) Although the hammer thrower has a much greater mass than the hammer, his moment of inertia about the axis is very much smaller than that of the hammer.
Explain this. A calculation could help your answer. [2]

(iii) The thrower releases the hammer after accelerating it from rest to 28 m s⁻¹ over 3.5 revolutions. Calculate the mean torque applied to the hammer. [4]

Option D – Energy and the environment

12. (a) A NASA website has the following statement:

> 'Averaged over the entire planet, roughly 340 watts per square meter of energy from the Sun reach Earth. About one-third of that energy is reflected back into space, and the remaining 240 watts per square meter is absorbed by land, ocean, and atmosphere.'

[Measuring Earth's Albedo (nasa.gov)]

(i) Explain how the 340 W m^{-2} figure is consistent with the published 1360 W m^{-2} value of the solar constant. [2]

....

....

....

(ii) (I) If only an average of 240 W reaches each square metre of surface directly from the Sun, calculate the additional power that must reach the Earth's surface per m^2 to achieve the mean equilibrium temperature of 16 °C and explain where this heat comes from. [4]

....

....

....

....

....

....

(II) The surface temperature of the Earth is rising because of human activities. Explain the role of greenhouse gases such as carbon dioxide (CO_2) in this. [3]

....

....

....

....

....

(b) Describe the process of *breeding* in the nuclear fuel cycle and state its importance. [4]

....

....

....

....

....

(c) Miriam is designing an industrial heat exchange system. It needs to raise the temperature of 2.0×10^{-3} m³ of water per second from 20 °C to 60 °C. The heat source is a tank of water at a constant temperature of 70 °C. A copper pipe conducts the cold water through the tank of hot water. This is Miriam's diagram:

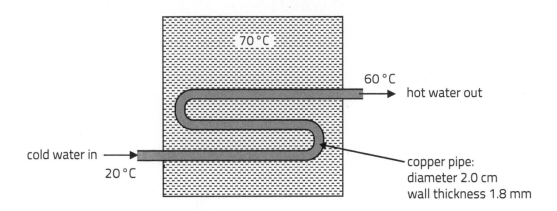

(i) Use the specific heat capacity equation:

$$Q = mc\Delta\theta$$

to show that the rate of heat transfer to the cold water is approximately 340 kW.   [3]

$[c_{water} = 4160 \text{ J kg}^{-1}\,°\text{C}^{-1};\ \rho_{water} = 1000 \text{ kg m}^{-3}]$

....

....

....

....

....

(ii) Miriam needs to know the length of copper pipe to use inside the tank. She makes the assumption that the mean temperature of the water inside the pipe is 40 °C.

Calculate the length of the pipe needed on the basis of Miriam's assumption.   [4]

$[K_{copper} = 385 \text{ W m}^{-1}\,°\text{C}^{-1}]$

....

....

....

....

....

....

....

Answers

Practice questions: Component 3: **Light, nuclei and options**

Section 1: The nature of waves

Q1 Waves involve no transfer of material; the particles in the medium just oscillate about a fixed point.

Q2 In a transverse wave, the particles of the medium oscillate at right angles to the direction of propagation. In a longitudinal wave, the particles of the medium oscillate parallel to the direction of propagation.
Examples: Transverse – seismic S waves; Longitudinal – seismic P waves.

Q3 (a) In a polarised light beam all the oscillations (of the electric field) are in the same direction.

(b) If a beam is partially polarised, oscillations occur in all directions at right angles to the direction of propagation but a fraction of the waves oscillate in the same direction.

Q4 (a) In an unpolarised beam, all directions have the same energy, so the energy in any component is 50% of the total.

(b) The unpolarised part has an intensity of 0.4 W m^{-2}, giving a constant intensity 0.2 W m^{-2}. The polarised part has an intensity of 0.6 W m^{-2}, so the transmitted intensity of this part should oscillate between 0 and 0.6 W m^{-2} with the maximum and minimum 90° apart. So the total intensity should oscillate between 0.2 and 0.8 W m^{-2}, 90° apart, which agrees with the graph and Cheryl is correct.

Q5

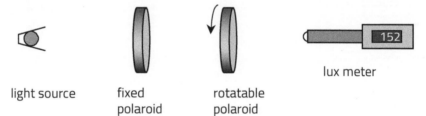

light source   fixed polaroid   rotatable polaroid   lux meter

The apparatus is set up as shown in the diagram. The rotatable polaroid is rotated through a set of positions, e.g. 15° apart. At each position the reading on the lux meter is noted and a graph plotted of reading against angle. The reading on the meter varies sinusoidally between a maximum and a minimum value. The angle between the two maxima is 180°. The graph is expected to look as follows.

Expected results:

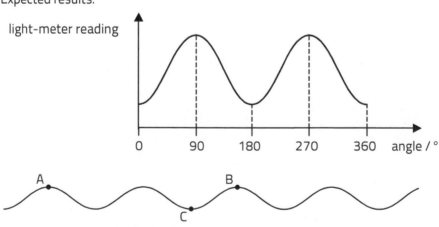

Q6 (a)

NB. In part (i), any peak; in part (ii), any trough.

(b) (i) $4\lambda = 3.04$ m, so $\lambda = 0.76$ m

(ii) Amplitude = 3.7 mm

(iii) $6T = 0.46$ s, so period = 0.0767 s
∴ speed = $\dfrac{0.76\,\text{m}}{0.0767\,\text{s}}$ = 9.9 m s^{-1}

Q7 (a)

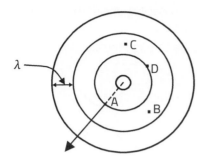

(b) Wavelength = $\dfrac{18.0\ \text{cm}}{3}$ = 6.0 cm; frequency = $\dfrac{25}{8.0\ \text{s}}$ = 3.125 Hz

Speed = λf = 6.0 × 3.125 = 19 cm s^{-1} (2 sf)

Q8 (a) P waves: $\lambda = \dfrac{v}{f} = \dfrac{6.2\ \text{km s}^{-1}}{8.9\ \text{Hz}}$ = 0.70 km [= 700 m] (2 sf)

S waves: $\lambda = \dfrac{v}{f} = \dfrac{3.7\ \text{km s}^{-1}}{8.9\ \text{Hz}}$ = 0.42 km [= 420 m] (2 sf)

(b) Let distance to Bala = d (in km)

Time for S waves from Liberty Stadium to Bala = $\dfrac{d}{3.7}$; time for P waves = $\dfrac{d}{6.2}$

∴ Time delay = $\dfrac{d}{3.7} - \dfrac{d}{6.2} = d\left(\dfrac{1}{3.7} - \dfrac{1}{6.2}\right)$ = 0.109d

∴ 0.109d = 16.3, ∴ $d = \dfrac{16.3}{0.109}$ = 150 km.

Q9 (a) The direction of travel is at right angles to the wavefronts. The direction of oscillation is at right angles to both the direction of travel and the wavefronts.

[Note: These ripples are actually 'surface waves' in which the motion of a particle is a vertical circle, with the plane of the circle in the direction of propagation. However, you will lose no marks if you treat them as transverse.]

(b) Number of wavelengths = 7.0, so $\lambda = \dfrac{14.6\ \text{cm}}{7}$ = 2.086 cm

$f = \dfrac{20}{4.7\ \text{s}}$ = 4.255 Hz

∴ Speed = λf = 2.086 cm × 4.255 Hz = 8.9 cm s^{-1}

(c) The number of waves passing any point in the shallow water must be the same as in the deep water (they aren't spontaneously created or destroyed), i.e. the frequency stays the same. The wavelength is given by $v = \lambda f$, so as the speed decreases, so does the wavelength and Gerallt is correct in both these statements.

Section 2: Wave properties

Q1 Diffraction is the spreading out of waves after they pass through a gap or past the edge of an obstacle.

Q2 When the slit is 300 nm wide, the waves spread out by 90° on both sides of the slit but the intensity of the transmitted light is very low. As the width increases (towards 600 nm) the intensity increases – more in the forward direction. Above 600 nm the transmitted waves become concentrated into an increasingly narrow central band with weaker side bands separated by narrow channels without any waves.

Q3 (a) The waves are produced by two dippers which vibrate in phase at the centres of the circles. The two sets of waves pass through each other. At points where the waves are in phase (e.g. at B), they add to give larger amplitude waves. At points where the waves are out of phase (e.g. at A and C) they subtract and cancel each other out.

(b) (i) 2λ

(ii) 1.5λ [or $1\frac{1}{2}\lambda$, if you prefer]

(c) There is a constant phase difference between the oscillations of the sources.

Q4 (a) The total displacement of two waves which pass through the same point is the [vector] sum of the displacements of the individual waves.

(b) (i)

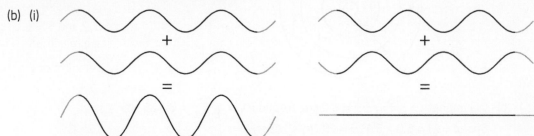

The left diagram shows constructive interference. The waves add to give one of double the amplitude. The right-hand diagram shows destructive interference. The waves add to give zero amplitude.

(ii)

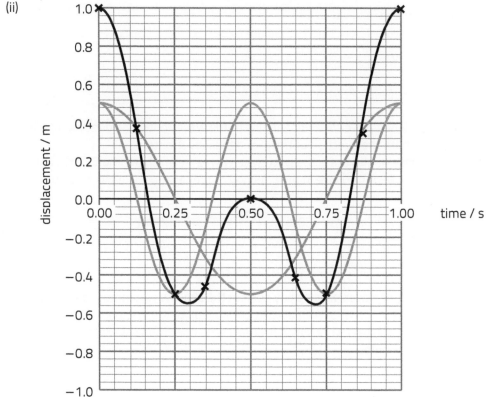

Q5 (a) It established that light was a wave phenomenon (not particles, as Newton had supposed).

(b)

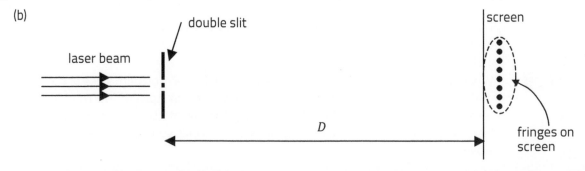

A blackened slide with two parallel slits about 0.5 mm apart is prepared and the separation, x, of the slits is measured using a travelling microscope. The slits are illuminated with a laser beam, as shown, and the fringes viewed on a distant screen ($D > 1$ m). The distance, D, to the screen is measured using metre rules or a tape measure. The separation, Δy, of the fringes is determined by measuring the distance between the outermost fringes using a mm scale and dividing by the number of fringe gaps visible. Δy is determined for a range of distances D, and a graph plotted of Δy against D.

The relationship between Δy and D is $\Delta y = \dfrac{\lambda D}{x}$, so the graph should be a straight line through the origin. The gradient $m = \dfrac{\lambda}{x}$, so the gradient, m, is determined and λ calculated using $\lambda = mx$.

Q6 (a) The microwaves incident on S_1 and S_2 spread out by diffraction and the divergent beams overlap at P and interfere. The waves are in phase at S_1 and S_2 and the distance S_1P and S_2P are equal, so the waves are in phase at P. Hence they interfere constructively, producing a maximum.

(b) For destructive interference, $S_2Q - S_1Q = \left(n + \frac{1}{2}\right)\lambda$ where λ is the wavelength of the microwaves and $n = 0, 1, 2.....$ As it is the first such point above P, $n = 0$ so $S_2Q - S_1Q = \frac{1}{2}\lambda$

For constructive interference, $S_2R - S_1R = n\lambda$. For R, $n = 1$ so, $S_2R - S_1R = \lambda$.

(c) (i) $\Delta y = \dfrac{\lambda D}{a}$, so PQ $= \dfrac{\lambda D}{2a} = \dfrac{2.8 \text{ cm} \times 15.0 \text{ cm}}{2 \times 6.5 \text{ cm}} = 3.23$ cm

(ii) Using Pythagoras, with PR = 6.5 cm:

$\dfrac{\lambda}{2} = S_2Q - S_1Q = \sqrt{15.0^2 + (3.23 + 3.25)^2} - \sqrt{15.0^2 + (3.23 - 3.25)^2} = 1.34$ cm

This gives a value of 2.7 cm for the wavelength, which is 4% too small – but still quite close!

Q7 (a)

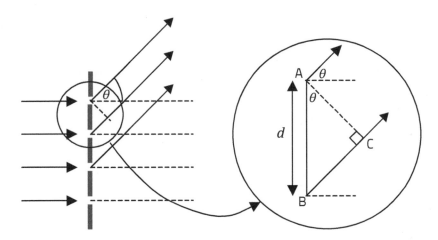

For the light from the top two slits to arrive at a distant point in phase, the path difference must be $n\lambda$ ($n = 0, 1, 2, ...$). On the expanded diagram, the angle BAC $= \theta$.

$\therefore$ Path difference = BC $= d \sin \theta$, where d = distance between slits.

So $n\lambda = d \sin \theta$.

Note: if this is true for two neighbouring slits, it is true for all slits.

(b) (i) When $n = 1$, $\theta = \tan^{-1}\left(\dfrac{0.225 \text{ m}}{1.750 \text{ m}}\right)$, $\therefore \sin \theta = 0.1275$

$d = \dfrac{1}{250 \times 10^3 \text{ m}^{-1}} = 4.00 \times 10^{-6}$ m

$\therefore \lambda = \dfrac{d \sin \theta}{n} = 4.00 \times 10^{-6} \times 0.1275 = 5.10 \times 10^{-7}$ m

(ii) Maximum possible value of $\sin \theta = 1.0$ $\therefore n_{\text{max}} = \dfrac{d}{\lambda} = \dfrac{4.00 \times 10^{-6} \text{ m}}{5.10 \times 10^{-7} \text{ m}} = 7.8$

n_{max} must be an integer, so possible values = 0, ±1, ±2, ...±7

$\therefore$ 15 bright dots.

Q8 (a) [see diagram right]

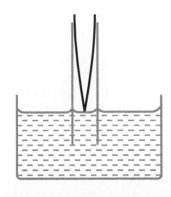

(b) Length $l = \dfrac{\lambda}{4}$, $\therefore \lambda = 4l$.

$c = \lambda f = 4lf$

(c) 317 m s^{-1}

309 m s^{-1}

All the values of c are lower than the true value ranging from 7% less at 256 Hz to 12% at 480 Hz. The % inaccuracy has a steady increasing trend as the frequency increases, suggesting a systematic uncertainty.

(d) $l = \dfrac{c}{4f}$, so 4 × gradient of a graph of l against $\dfrac{1}{f}$ should be c.

No grid available, so taking the extreme values: when $l = 0.312$ m, $\dfrac{1}{f} = 3.91 \times 10^{-3}$ s

When $l = 0.157$ m, $\dfrac{1}{f} = 2.08 \times 10^{-3}$ s

$\therefore$ 4 × gradient = $4 \times \dfrac{(0.312 - 0.157)\,\text{m}}{(3.91 - 2.08) \times 10^{-3}\,\text{s}} = 339$ m s^{-1}, which is within 1% of the true value.

Section 3: Refraction of light

Q1 (a) At the boundary between two given materials, the ratio of the sine of the angle of incidence to the sine of the angle of refraction is a constant.

(b) For light passing from a vacuum into the material, $n = \dfrac{\sin i}{\sin r}$ where i is the angle of incidence and r is the angle of refraction.

(c) $n = \dfrac{c}{v}$, where v is the speed of light in the material and c the speed of light in a vacuum.

Q2 $n = \dfrac{c}{v}$, so $v = \dfrac{c}{n} = \dfrac{3.00 \times 10^8}{1.49}$ m s^{-1} = 2.01×10^8 m s^{-1}

Q3 (a) Because electrons cannot go faster than the speed of light in a vacuum.

(b) (i) Speed of light in water = $\dfrac{c}{n} = \dfrac{3.00 \times 10^8}{1.33} = 2.26 \times 10^8$ m s^{-1}. Any faster than this and Cherenkov radiation will be produced.

(ii) Using $\tfrac{1}{2}mv^2 = eV$, $V = \dfrac{mv^2}{2e} = \dfrac{9.11 \times 10^{-31} \times (2.26 \times 10^8)^2}{2 \times 1.60 \times 10^{-19}} = 145$ kV

Q4 (a) $v_{\text{air}} > v_{\text{water}} > v_{\text{glass}}$. Hence the angle, θ, to the normal in the three materials follows the same pattern: $\theta_{\text{air}} > \theta_{\text{water}} > \theta_{\text{glass}}$.

(b) $1.00 \sin 45° = 1.52 \sin x \therefore x = \sin^{-1}\left(\dfrac{\sin 45°}{1.52}\right) = \sin^{-1} 0.4652 = 27.7°$

$y = x$ [alternate angles] = 27.7°.

$1.52 \sin 27.7° = 1.33 \sin z$, $\therefore z = \sin^{-1}\left(\dfrac{1.52 \sin 27.7°}{1.33}\right) = 32.1°$

(c) $n \sin \theta$ is constant, so $1.00 \sin 45° = 1.33 \sin z$ and the glass is irrelevant (apart from holding the water in place).

Q5 (a) $r = \sin^{-1}\left(\dfrac{\sin 27°}{1.42}\right) = 18.6°$

(b) Angle of incidence on back surface = 18.6° (angles in a segment) so angle of refraction back into air = 27°.

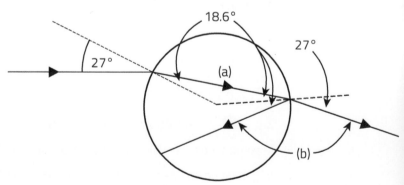

Q6 (a) Light travelling in a medium is incident on another medium, in which the speed of light is higher, with an angle of incidence greater than the critical angle.

(b) Critical angle, $c = \sin^{-1}\left(\dfrac{1}{n}\right) = \sin^{-1}\left(\dfrac{1}{1.55}\right) = 40.2°$.

(c) Angle of refraction = 19.8° (by considering angles in the triangle)
$\therefore i = \sin^{-1}(1.55 \sin 19.8°) = 31.7°$

(d) The angle of incidence on the bottom surface is 19.8° (see diagram), which is less than the critical angle so total internal reflection does not occur and some of the light will emerge and Briony is incorrect. In fact the light emerges at the same angle as it entered the prism (31.7°).

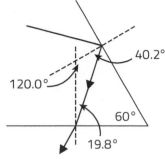

Q7

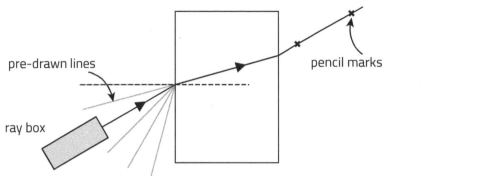

pre-drawn lines

pencil marks

ray box

Place the glass block in the middle of a large piece of paper and draw its outline. Mark a series of lines at a regular set of angles to the normal (e.g. 15°, 30°, 45°, 60°, 75°) as shown and shine a narrow beam from a ray box along one of them. Put marks at well-separated points on the emergent ray as shown. Repeat for each of the other pre-drawn line. Remove the block and use the outline, the pencil marks and a ruler to reconstruct each of the paths for the light through the block. For each line, use a protractor to measure the angles of incidence, i, and refraction, r, at the point of incidence.

Plot $\sin i$ against $\sin r$ and draw a best-fit line, which should be straight and through the origin. Determine the gradient of the line – this is the refractive index.

Q8
When the level of the benzene is as shown, the ray of light passes straight through the prism into the benzene, almost undeviated, because the refractive indices of the benzene and glass are almost the same. If the level of benzene drops below the 'minimum height' mark the light ray is totally reflected on both lower surfaces (first horizontal and then vertical) and emerges back out and is detected by the alarm. This occurs because the angle of incidence (45°) is greater than the critical angle for the glass: $c = \sin^{-1}\left(\frac{1}{1.5}\right) = 42°$.

Q9 (a) $\theta = \sin^{-1}\left(\frac{1.00 \times \sin 21.2°}{1.63}\right) = 12.8°$

$\phi = 90° - \theta = 77.2°$.

(b) The critical angle for the boundary between the fibre core and the cladding is given by:

$c = \sin^{-1}\left(\frac{1.60}{1.63}\right) = 79.0°$

The angle of incidence is less than the critical angle so total internal reflection does not occur.

(c) (i) Distance travelled $= \frac{14.0\,\text{km}}{\cos 5°} = 14.053$ km. So the extra distance = 53 m.

(ii) The extra distance of 53 m takes light a time of $\frac{53\,\text{m}}{1.84 \times 10^8\,\text{m s}^{-1}} = 2.9 \times 10^{-7}$ s.

The signal consists of a series of pulses of light. So for pulse intervals greater than about 10^{-7} s (i.e. a pulse frequency of 10 MHz) the pulses which travel by different paths will overlap with earlier or later pulses and the pulses will be unreadable – this is multi-mode dispersion.

Q10

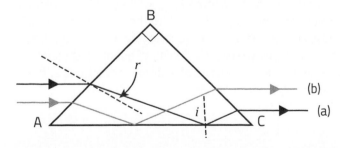

(a) The angle of refraction, $r = \sin^{-1}\left(\dfrac{1.00 \times \sin 45°}{1.60}\right) = 26.2°$. The angle of incidence, i on AC is 71.2°, which is greater than the critical angle and hence TIR occurs. The angle of incidence on BC is 26.2° so the angle of refraction is 45°, i.e. the emergent ray is parallel to the initial ray.

(b) (See grey rays on diagram.) The angles are the same so the light rays are parallel but the upper one on the left is the lower one on the right.

(c) If the refractive index were, say, 1.3, the angle of incidence on AC would be 77° so, yes there would be TIR if the light ray hits AC. However, the lower the refractive index, the larger the angle of refraction on AB and if it is large enough, the ray wouldn't hit AC at all. So James is right, but for low refractive indices it will only work for light rays that hit AB near the bottom.

Section 4: Photons

Q1 (a) A photon is a particle of light. [It has energy hf where h is the Planck constant.]

(b) Photon energy = hf.
If there are n photons per second, energy transfer per second = nhf.

Q2 A line emission spectrum consists of a series of narrow bands of light, each with a particular frequency / colour / wavelength. It is produced by hot, low-pressure, gases which consist of individual atoms and/or molecules. It can be displayed by passing the light through a diffraction grating or a prism and projecting the resulting beams onto a screen.

Q3 Light from the surface [photosphere] passes through a tenuous atmosphere of gas. Gas atoms absorb photons of light at specific wavelengths, corresponding to the difference in energy levels of the atoms. The excited atoms emit photons of the same wavelength but in random directions, so that if the spectrum of the light is observed from the Earth it appears as the continuous spectrum of light from the photosphere with a series of dark lines across it – the absorption spectrum.

Q4 (a) De Broglie wavelength, $\lambda = \dfrac{h}{p}$. The momentum, $p = \sqrt{2mE} = \sqrt{2meV}$

So $\lambda = \dfrac{h}{\sqrt{2meV}} = \dfrac{6.63 \times 10^{-34}}{\sqrt{2 \times 9.11 \times 10^{-31} \times 1.60 \times 10^{-19} \times 2400}} = 2.51 \times 10^{-11}$ m

(b) From the calculation in part (a) we see that as the voltage is increased, the wavelength of the electrons decreases. Hence the diffraction pattern gets smaller [the sine of the angle to the centre of the diffracted beam is proportional to the wavelength].

Q5 (a) (i) The photon energy is the difference in the two energy levels.
So energy = 12.1 eV – 10.2 eV = 1.9 eV

(ii) $E = hf = h\dfrac{c}{\lambda}$

So $\lambda = \dfrac{hc}{E} = \dfrac{6.63 \times 10^{-34} \times 3.00 \times 10^{8}}{1.9 \times 1.60 \times 10^{-19}} = 6.5 \times 10^{-7}$ m; visible (red)

(b) (i) Greatest energy = 12.75 – 0.0 = 12.75 eV
$= 12.75 \times 1.60 \times 10^{-19}$ J
$= 2.0 \times 10^{-18}$ J

(ii) Ultra-violet.

Q6 (a) Einstein's photoelectric equation can be written $E_{k\,max} = hf - \phi$. It is an expression of conservation of energy: the maximum electron energy $E_{k\,max}$ is equal to the photon energy (hf) minus the minimum energy needed to remove an electron from the surface (the work function, ϕ). It provided a firm experimental foundation for the wave–particle duality of quantum theory by showing that light has properties that are only explicable if it consists of a stream of particles with energy proportional to the frequency (a wave property).

To obtain the results, the photocell is illuminated with monochromatic radiation (of high enough frequency) and the supply pd increased until the photo-current just becomes zero. This pd value is multiplied by e (the electronic charge) to give $E_{k\,max}$. This is repeated for several frequencies and the graph drawn.

(b) All photons of a monochromatic beam have the same energy. The intensity of the beam is proportional to the number of photons per second passing a point [incident on the emitting surface]. Each photon has the same probability of causing the emission of an electron, so the number of electrons emitted per second and received on the collecting electrode is proportional to the intensity. Hence the current is also proportional.

Q7 (a) The photons in the beam each have a momentum, p, given by the de Broglie relationship, $p = h/\lambda$. As each is reflected from the mirror, it suffers a change in momentum. The space probe receives an equal and opposite change of momentum, i.e. a force.

(b) (i) $p = \dfrac{h}{\lambda}$. Also $E = hf = \dfrac{hc}{\lambda}$.

Substituting for $\lambda \longrightarrow p = h \times \dfrac{1}{\lambda} = h \times \dfrac{E}{hc}$, so $E = pc$.

(ii) Change in momentum per second of each photon $= 2 \times \dfrac{E}{c}$ because of the reflection.

So force on probe $= 2\dfrac{E}{c} \times$ number of photons per second $= 2\dfrac{P}{c} = 2 \times \dfrac{15\,000}{3.00 \times 10^8}$ N

$= 0.10$ mN

So $a = \dfrac{F}{m} = \dfrac{0.1 \times 10^{-3}}{2.3} = 4.3 \times 10^{-5}$ m s^{-2}.

(iii) (I) $v = u + at$, so speed $= 4.3 \times 10^{-5} \times (86\,400 \times 365) = 1400$ m s^{-1}

(II) $x = ut + \frac{1}{2}at^2 = 0.5 \times 4.3 \times 10^{-5} \times (86\,400 \times 365)^2$

$= 2.1 \times 10^{10}$ m [21 million km]

(iv) The laser is aimed at the mirror which reflects the beam back towards the probe, thus producing a reverse thrust.

(c) (i) Assume that the diameter of the window is 2 mm.

Then $\theta = \dfrac{2\lambda}{d} = \dfrac{2 \times 500 \times 10^{-9}\,\text{m}}{2 \times 10^{-3}\text{m}} = 5 \times 10^{-4}$ rad.

So, from the back of the lecture room to the front, the diameter, D, of the circle of the spot is given by:
$D = 5 \times 10^{-4} \times 10$ m $= 5 \times 10^{-3}$ m $= 5$ mm.
A spot of light of diameter 5 mm is small enough to be useful as a pointer, so diffraction is not a problem.

(ii) [There are many ways of answering this AO3 question. This is one of them.]
If the power incident on the spacecraft is 10% of the original, then the area of the beam is 10 × the area of the receiving surface, so the diameter is $\sqrt{10} \times 100$ m $= 320$ m.

The diffraction angle $\theta = \dfrac{2\lambda}{d} = \dfrac{2 \times 500 \times 10^{-9}\,\text{m}}{1\,\text{m}} = 1 \times 10^{-6}$ rad.

So distance for the diameter of beam to be 320 m $= \dfrac{320\,\text{m}}{1 \times 10^{-6}\text{rad}} = 3.2 \times 10^8$ m

This distance, 300 000 km is much less than the distance travelled in 1 year calculated in part (b), so the speed achieved will also be much less and the spacecraft will take many years to reach Mars.

Section 5: Lasers

Q1 An atom is in an excited state. A photon with energy equal to the energy difference between the excited state and an unoccupied lower energy state triggers the atom to change to a lower energy state. This releases a second photon – *stimulated emission*.

Q2 (a) If there are two energy states in a system (e.g. a collection of atoms) the lower energy state usually has a greater population. A population inversion is when there are more atoms in the higher energy state than in the lower.

(b) An incident photon (of the correct energy) can be absorbed and cause an atom in the lower energy state to jump to the higher. An identical photon can cause an atom in the higher energy state to fall to the lower. In a population with equal numbers in the lower and higher energy states, the optical pumping will cause equal numbers of transitions in the two directions, so the number of atoms in the higher energy state cannot increase above this level.

Q3

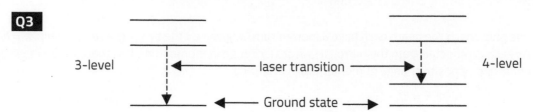

3-level   laser transition   4-level

Ground state

In a 3-level system the lower level in the laser transition is the ground state. So, in order to get a population inversion with an excited state, at least half the atoms must be in the upper level. In a 4-level system, the lower level in the laser transition is a normally empty excited level, so only a small proportion of atoms need to be put into the upper level, which is easier to achieve.

Q4 The pumping level needs to empty quickly so:

(i) the population in the upper level builds up quickly, and

(ii) to reduce stimulated emission between the pumping and ground levels [OR, equivalently, to stop the top level filling up (making pumping less efficient)].

Q5 Uses: in optical fibres and DVD reading heads
Advantages: small, efficient (low power use) [also very cheap]

Q6 (a) Laser transition $E_3 \longrightarrow E_2$,
so energy = 19.5 eV − 17.4 eV = 2.1 eV = 2.1 × 1.60 × 10⁻¹⁹ = 3.36 × 10⁻¹⁹ J
$$f = \frac{E}{h} = \frac{3.36 \times 10^{-19}\,\text{J}}{6.63 \times 10^{-34}\,\text{J s}} = 5.1 \times 10^{14}\,\text{Hz (2 sf)}$$

(b) The transition is stimulated by a photon of energy 2.1 eV. The system is chosen so that level E_3 is a metastable energy level, i.e. it has a long lifetime, so the probability of a spontaneous emission is low. However, a photon of energy 2.1 eV has a high probability of triggering emission. [Also, there is so much light inside a laser that the photon will be stimulated long before spontaneous emission is likely to occur.]

(c) To produce each laser photon of energy 2.1 eV, the pumping energy required is 20.5 eV. Some of the pumped energy will be wasted by a transition back down to the ground state.
Hence the maximum efficiency = $\frac{2.1\,\text{eV}}{20.5\,\text{eV}} \times 100\% = 10.2\%$. So Joel is correct to 2 sf [and some energy will inevitably be lost anyway].

(d) In order to establish a population inversion between E_2 and E_3, the population of E_2 must be kept as low as possible. Hence the second energy level must have a low lifetime, so that it can empty back down to the ground state as quickly as possible and Nigella is wrong.

Q7 (a) Pumping energy = 4.8 × 10⁻¹⁹ J = $\frac{4.8 \times 10^{-19}}{1.60 \times 10^{-19}}$ eV = 3.0 eV

Violet [but a marking scheme would accept blue, indigo or violet].

(b) E_3 the pumped level should have a short lifetime and E_2 a much longer one. This is to allow a population inversion to be established between E_2 and E_1.

(c) $E = hf$ and $c = f\lambda$ ∴ $\lambda = \frac{hc}{E} = \frac{6.63 \times 10^{-34} \times 3.00 \times 10^8}{3.1 \times 10^{-19}} = 6.4 \times 10^{-7}\,\text{m}$

(d) Same frequency [or wavelength or energy]
Same phase
Same direction
Same polarisation [NB only 3 asked for – don't give 4 in case one is wrong!]

(e) Paula is correct. To achieve a population inversion between E_1 and E_2, the population of E_2 (i.e. N_2) must be greater than N_1. This happens most efficiently if all the pumped atoms immediately drop down to E_2. N_1 is more than 50% of the total, then N_2 must be less and a population inversion doesn't exist.

Q8 (a) The exiting laser beam removes some photons from the cavity, so to keep the number as high as possible, there should be no losses at the left hand mirror.

(b) Some photons need to escape to form the exiting laser beam.

(c) (i) More photons (90%) are reflected on the left-hand mirror than on the right (40%), so there is more momentum change of the photons per second on the left than the right. Hence, by Newton's second law, there is a greater force on the left.

(ii) 60% of photons incident on right-hand mirror escape.

$\therefore$ Power of photons incident on mirror $= \dfrac{5.0\text{ mW}}{0.6} = 8.33$ mW

$\therefore$ Power reflected = 8.33 mW – 5.0 mW = 2.33 mW

$\therefore$ Force exerted $= 2 \times \dfrac{P}{c} = 1.55 \times 10^{-11}$ N

$\therefore$ Pressure $= \dfrac{F}{A} = \dfrac{1.55 \times 10^{-11}\text{N}}{0.94 \times 10^{-6}\text{ m}^2} = 1.7 \times 10^{-5}$ Pa (2 sf)

(d) Consider N photons being reflected from right mirror. Assume number hitting left-hand mirror = kN, where k is the amplification factor for one pass. The number reflected from left-hand mirror = kN. Number hitting right-hand mirror = k^2N. Then number reflected from right-hand mirror = $0.95 \times k^2N$.

$\therefore$ To achieve equilibrium $0.95k^2 N = N$

$\therefore k = \dfrac{1}{\sqrt{0.95}} = 1.026$, so the increase is 2.6% and Helena is correct – she's a bright cookie!

Section 6: Nuclear decay

Q1 The β particles could be <u>positrons</u> (β⁺ radiation). The particles are of <u>high energy</u> and originate in the <u>unstable nuclei</u> of certain atoms.

Q2 (a) The activity of a radioactive nuclide is the number of disintegrations per unit time.
Unit: becquerel (Bq) which is equivalent to s^{-1}.

(b) 1 year $= 60 \times 60 \times 24 \times 365$ s $= 3.15 \times 10^7$ s
If there are N atoms, no. of decays in 1 year $= 9.11 \times 10^{-13} \times 3.15 \times 10^7\ N$
$= 2.87 \times 10^{-5}\ N$

$\therefore$ Probability of a particular atom decaying $= 2.87 \times 10^{-5} = \dfrac{1}{34\,800}$

$\therefore$ The true probability is 1 in 34 700 which is indeed less than 1 in 30 000.

Q3 (a) All the readings are the same to within the expected variability of the random emissions. Any α radiation from the sample would not penetrate 10 cm of air, so the experiment gives no information about α. The lack of a reduction with the thin aluminium suggests that the sample is a γ emitter.

(b) A background reading should be taken to establish the readings from the source alone. The readings should be repeated at 2 cm distance to allow any α emissions to be detected.

Q4 (a) $^{235}_{92}\text{U} \rightarrow {}^{231}_{90}\text{Th} + {}^{4}_{2}\text{He}$ [or ${}^{4}_{2}\alpha$]

(b) (i) The mass number (nucleon number), A, must be $235 - 4n$ where n is an integer because the ${}^{4}_{2}\text{He}$ has $A = 4$. $207 = 235 - 4 \times 7$, so ${}^{207}_{82}\text{Pb}$ is the end of the decay series.

(ii) $206 = 238 - 4 \times 8$, so there are 8 alpha decays. Without β^- this would produce a nuclide with $Z = 92 - 2 \times 8 \doteq 76$. So 6 β^{-1} decays are needed to give $Z = 82$.

$$^{238}_{92}\text{U} \rightarrow {}^{206}_{82}\text{Th} + 8\,{}^{4}_{2}\text{He} + 6\,{}^{0}_{-1}\text{e}$$

(iii) There is no whole number n for which $233 - 4n$ is equal to 206, 207 or 208, so any isotope of lead produced, e.g. 205, would be unstable and cannot be the end product.

Q5

Method 1: equal times → equal ratios

(a) (i) After 1 year $A \rightarrow \dfrac{9.76}{11.50} A = 0.849A$.

∴ In another year, the count rate will be $0.849 \times 9.76 = 8.28$

(ii) Count rate $= (0.849)^{10} \times 9.76 = 1.90$ count per second.

(b) In n years, count rate $= 9.76(0.849)^n$.

If $9.76(0.849)^n = 0.42$, taking logs: $\ln 9.76 + n \ln 0.849 = \ln 0.42$

∴ $n = \dfrac{\ln 0.42 - \ln 9.76}{\ln 0.849} = 19.2$ years (3 sf)

Method 2: calculating the decay constant

(a) (i) After 1.00 year, $9.76 = 11.50e^{-1.00\lambda}$, so $\ln 9.76 = \ln 11.50 - \lambda \rightarrow \lambda = 0.164$ year^{-1}

∴ In another year, the $C = 9.76e^{-0.164 \times 1.00} = 8.28$

(ii) $C = 9.76e^{-0.164 \times 10.00} = 1.89$ count per second.

(b) In n years $C = 9.76e^{-0.164n}$.

If $9.76e^{-0.164n} = 0.42$, taking logs: $\ln 9.76 - 0.164n = \ln 0.42$

∴ $n = \dfrac{\ln 9.76 - \ln 0.42}{0.164} = 19.2$ years (3 sf)

Method 3: Using half-lives $C = C_0\,2^{-n}$ [left for you]

Q6 (a) ${}^{1}_{0}\text{n} + {}^{14}_{7}\text{N} \rightarrow {}^{14}_{6}\text{C} + {}^{1}_{1}\text{H}$

(b) ${}^{14}_{6}\text{C} \rightarrow {}^{14}_{7}\text{N} + {}^{0}_{-1}\text{e} + {}^{0}_{0}\overline{\nu}_e$

(c) (i) $\lambda = \dfrac{\ln 2}{5730\ \text{year}} = 1.210 \times 10^{-4}$ year^{-1} [$= 3.83 \times 10^{-12}$ s^{-1}]

(ii) Minimum age: $e^{-\lambda t} = \dfrac{0.853}{1.250} = 0.6824$, ∴ $-1.210 \times 10^{-4}\,t = \ln 0.6824$

∴ $t_{min} = 3158$ year.

Maximum age: $e^{-\lambda t} = \dfrac{0.849}{1.250} = 0.6792$, ∴ $-1.210 \times 10^{-4}\,t = \ln 0.6792$

∴ $t_{max} = 3197$ year. ∴ Age $= 3180 \pm 20$ year

(iii) **Qualitative answer:** The fraction of ^{14}C in modern objects is less that it would be without the addition of the fossil fuels. Objects which are old also have a lower fraction of ^{14}C, which is the same effect, so Sioned is correct.

Quantitative answer: If apparent age of modern object $= t$, $e^{-\lambda t} = 0.97$

∴ $t = \dfrac{\ln 0.97}{-1.21 \times 10^{-4}} = 250$ years. So modern objects appear 250 years old and Sioned is correct.

Q7 (a) With 8 faces, probability of remaining after 1 throw $= \dfrac{7}{8} = 0.875$.

∴ No. remaining $= 800 \times 0.875$ ∴ After n throws $800 \times (0.875)^n$

(b) For half to remain, $(0.875)^n = 0.5$,

∴ $n = \dfrac{\ln 0.5}{\ln 0.875} = 5.19$ throws. The points agree with the theoretical curve and half-life.

(c) Theoretical no. remaining = $800 \times (0.75)^n \rightarrow$ In 2-throw steps: 450, 253, 142, 80, 45, 25

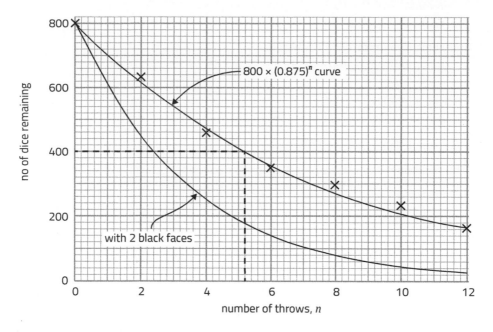

Q8 (a) Range of 1.0 MeV β particles = 0.41 cm

$$\therefore \text{Range in glass} = \frac{1.0 \times 10^3 \text{ kg m}^{-3}}{2.5 \times 10^3 \text{ kg m}^{-3}} \times 4.1 \text{ cm} = 1.6(4) \text{ cm}$$

(b) The range of 2.0 MeV β particles is 0.97 cm, showing that they lose the first MeV in 0.56 cm (i.e. 0.97 cm − 0.41 cm) and the last MeV in 0.41 cm, hence Dylan is correct.

(c) (i) $^6_3\text{Li} + ^1_0\text{n} \rightarrow ^3_1\text{H} + ^4_2\text{He}$

(ii) $^3_1\text{H} \rightarrow ^3_2\text{He} + ^0_{-1}\text{e} + ^0_0\overline{\nu}_e$

(iii) The range of 0.1 MeV β particles is about 0.01 cm in water and therefore about 0.04 mm in glass. Hence, the β particles cannot penetrate the walls of the tubes.

Q9 (a)

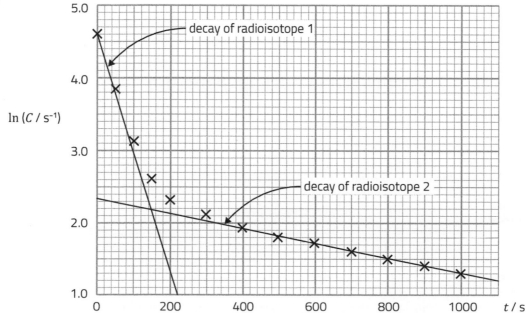

The ln C vs. t graph consists of two straight line portions of different negative gradients. [Note: The curved part represents the period in which both isotopes contribute similarly to the counts.]

(b) (i) Decay constant = − gradient = $-\dfrac{1.19 - 2.33}{1100 \text{ s}}$

$\therefore \lambda = 1.04 \times 10^{-3} \text{ s}^{-1}$

(ii) For isotope 2, at $t = 0$, $\ln(C/s^{-1}) = 2.33$, where C is the count rate
∴ $C / s^{-1} = e^{2.33} = 10.3$
i.e. $C = 10.3\ s^{-1}$

(c) Initial total count rate $= e^{4.6} = 99.5\ s^{-1}$
∴ Count rate from isotope 1 $= 99.5 - 10.3 = 89.2\ s^{-1}$

Q10 (a) $^{238}_{92}U \rightarrow\ ^{234}_{90}Th + ^{4}_{2}He$
$^{234}_{90}Th \rightarrow\ ^{234}_{91}Pa + ^{0}_{-1}e + ^{0}_{0}\overline{\nu_e}$

(b) After n half-lives $C = C_0 \times 2^{-n}$ ∴ $n = \dfrac{\ln(C_0/C)}{\ln 2}$
The maximum decay is $492 \rightarrow 77$ and the minimum decay is $448 \rightarrow 95$ counts
In 3 minutes: $n_{max} = \dfrac{\ln(492/77)}{\ln 2} = 2.68$,
∴ $t_{1/2\ min} = \dfrac{3.0\ \text{minutes}}{2.68} = 1.12\ \text{minutes}$
and $n_{min} = \dfrac{\ln(448/95)}{\ln 2} = 2.23$,
∴ $t_{1/2\ max} = \dfrac{3.0\ \text{minutes}}{2.23} = 1.34\ \text{minutes}$
The figure of 1.17 minutes is within the experimental range, so the student's results are consistent with it.

(c) Repeat the experiment several times and measure the 10 second count rates C, more frequently, e.g. every 20 s. Add the results for each time together and plot a graph of $\ln C$ against t. The straight-line graph has a gradient of $-\lambda$ from which the half-life can be calculated using $t_{1/2} = (\ln 2) / \lambda$.

Q11 (a) A charged particle, q, moving with a velocity, v, across a magnetic field, B, experiences a force, F, at right angles to B and v in the direction given by Fleming's Left Hand Motor rule. If q is negative, the direction of F is opposite. In this case, the charges are accelerated downwards in the field showing them to be negative (β^-).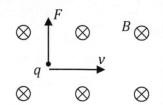

(b) Photons are uncharged and so will be undeflected, i.e. they pass through in a straight line. α particles are positively charged, so experience an upward force; however, they have a much greater mass than β particles so the deflection is very small and would not be observed with this simple arrangement.

Section 7: Particles and nuclear structure

Q1 Electrons have no structure – they cannot be separated into constituent particles.
Protons are composed of 3 quarks: 2 up quarks and 1 down quark.

Q2 The particle is a hadron: it is composed of quarks and/or anti-quarks

Q3 Strong: proton, pi+ meson, anti-neutron
Electromagnetic: electron, proton, pi+ meson, anti-neutron, positron
Weak: all of them

Q4 (a) Electromagnetic

(b) (i) The decay time is appropriate for e-m interaction (intermediate)

(ii) Photons are produced (so cannot be strong)

Q5 (a) (i) positron, e^+

(ii) anti-neutron, $\overline{n}$

(iii) anti-electron neutrino (or electron anti-neutrino), $\overline{\nu_e}$

(b) The π^- has quark structure $d\bar{u}$; the π^+ has structure $u\bar{d}$. So the quarks in π^+ are the anti-quarks to those in π^-, so Eurig is correct.

Q6 (a) It cannot be a lepton because there are no leptons with $Q = 2$. A hadron with $Q = 2$ must have 3 quarks because no combination of a quark and anti-quark, which all have charges $\pm\frac{1}{3}$ or $\pm\frac{2}{3}$, can have $Q = 2$. So it must be a baryon.

(b) (i)   Strong interaction because of the short time-scale

(ii) Baryon number, charge, quark flavour, lepton number

(iii) $\Delta^{++} \longrightarrow p + \pi^+$. At the quark level, this is $uuu \longrightarrow uud + u\bar{d}$
Charge: $Q(\Delta^{++}) = 2$; $Q(p) + Q(\pi^+) = 1 + 1 = 2$, $\therefore$ conserved.
Baryon number: $B(\Delta^{++}) = 1$; $B(p) + B(\pi^+) = 1 + 0 = 1$, $\therefore$ conserved.
Quark flavours: $U(\Delta^{++}) = 3$; $U(p) + U(\pi^+) = 2 + 1 = 3$ $\therefore$ conserved.
$D(\Delta^{++}) = 0$; $D(p) + D(\pi^+) = 1 + (-1) = 0$ $\therefore$ conserved
Lepton number: no leptons involved, so $L = 0$ on left and right, $\therefore$ conserved

Q7 (a) π^+ is a meson; e^+ is a lepton; v_e is a lepton

(b) Charge is conserved (the pion and positron are both +; the neutrino is neutral)
Lepton number is conserved: $L(\pi^+) = 0$; $L(e^+) + L(v_e) = -1 + 1 = 0$
Baryon number is conserved: 0 for all particles

(c) Quark flavour is not conserved: For the π^+, $U = 1$, $D = -1$. For the positron and neutrino both U and D are zero.

Q8 (a) Charge: the left-hand side has $Q = 1 + 1 = 2$; right-hand side $Q = = 1 + (-1) + 1 = 1$
Baryon number: $B(p + p) = 1 + 1 = 2$; $B(\Delta^+ + e^- + \pi^+) = 1 + 0 + 0 = 1$
Up number: $U(p + p) = 2 + 2 = 4$; $B(\Delta^+ + e^- + \pi^+) = 2 + 0 + 1 = 3$
Down number: $D(p + p) = 1 + 1 = 2$; $D(\Delta^+ + e^- + \pi^+) = 1 + 0 + (-1) = 0$
Lepton number: $L(p + p) = 0 + 0 = 0$; $L(\Delta^+ + e^- + \pi^+) = 0 + 1 + 0 = 1$

(b) This decay, producing a photon, is an electromagnetic decay, which is slower than the first decay, which is a strong decay.

Q9 Lepton number $L(n + \pi^+) = 0$; $L(\Delta^{++} + e^-) = 0 + 1 = 1$
Charge: $Q(n + \pi^+) = 0 + 1 = 1$; $Q(\Delta^{++} + e^{-1}) = 2 - 1 = 1$. So not violated.
Baryon number: $B(n + \pi^+) = 1 + 0 = 1$; $Q(\Delta^{++} + e^-) = 1 - 0 = 1$. So not violated.

Q10 (a) Baryons have a baryon number of 1. 3 quarks have a baryon number of $\frac{1}{3} + \frac{1}{3} + \frac{1}{3} = 1$, e.g. the neutron has quark structure udd.

(b) Mesons each have a quark, with baryon number $\frac{1}{3}$, and an antiquark with baron number $-\frac{1}{3}$, e.g. the pion, π^+, has quark structure $u\bar{d}$. So the total baryon number is $\frac{1}{3} - \frac{1}{3}$, i.e. 0.

Q11 The protons and electrons can interact via the electromagnetic interaction, which has a long range. They can lose energy to electrons in the atoms through which they pass. The protons can also interact with nucleons in the lead nuclei via the strong force, but this has a minor effect as the range of the force is only about 10^{-15} m.
The neutrinos can interact with electrons and nucleons only with the weak force. Hence they have to approach within $\sim 10^{-17}$ m and the probability of an interaction is very small.

Q12 **Lepton number.** Leptons are fundamental particles. The lepton family consists of 3 generations. The first generation members are the electron and the [electron] neutrino. They each have a lepton number, L, of 1. The antiparticles to these (the positron and the anti-neutrino) have lepton numbers of −1. In any interaction, the total lepton number is conserved, e.g. if an electron is the only reacting particle, either an electron or a neutrino must be a product particle.

Baryon number. Baryons comprise 3 quarks, which are fundamental particles, and each has a baryon number, B, of 1. Anti-baryons, comprising 3 anti-quarks, have $B = -1$. In any interaction the baryon number is conserved.

Charge: This is always conserved in particle interactions.

Quark flavour: First-generation quarks have two flavours, up (u) and down (d), each with a separate flavour number, U and D. U and D are conserved in strong and electromagnetic interactions but may change by ± 1 in weak interactions.

Q13 (a) The decay of a $\Delta^+ \longrightarrow p + \pi^0$ conserves charge ($Q = 1$), baryon number ($B = 1$) and lepton number ($L = 0$) and the mass of the Δ^+ is greater than p + π^0 together.
The mass of p + ρ^0 ($3352m_e$) is greater than the mass of the Δ^+.

(b) In the decay $\pi^0 \longrightarrow e^- + e^+ + \gamma$, the charge, baryon number and lepton number are all conserved – they are all 0 on both sides of the equation. Hence the decay is possible (by the electromagnetic interaction). The mass of the product particles ($2m_e$) is less than the mass of the pion.

(c) Baryon number must be conserved. The only baryon less massive than the neutron is the proton, so the neutron must decay into a proton.
The reaction n $\longrightarrow$ p + π^- is impossible because the total mass of p + π^- ($2043m_e$) is greater than that of the neutron.

So the neutron can only decay into a proton and an electron ($m = 1837m_e$). In order to conserve lepton number, an electron anti-neutrino must also be produced. Hence it must be a weak decay.

(d) To preserve baryon number, it would have to decay into another baryon but the proton is the lightest baryon, so this is impossible.

Section 8: Nuclear energy

Q1 (a) The 'energy to hold the particles together' would have mass, given by $E = mc^2$. So the total mass would be the mass of the protons + the mass of the neutrons + the mass of this energy.

(b) Binding energy is the energy required to separate the component nucleons.

Q2 13.6 eV = 2.18×10^{-18} J.
This energy has mass = $\dfrac{2.18 \times 10^{-18} \text{ J}}{(3.00 \times 10^8 \text{ m s}^{-1})^2}$ = 2.42×10^{-35} kg = 1.46×10^{-8} u.
So the mass of the atom is 0.000 000 015 u less than the sum of the masses of the proton and the electron. Hence they are different to this number of significant figures.

Q3 (a) Mass deficit = $2 \times (1.007\,276 + 1.008\,665 + 0.000\,549) - 4.002\,604$ u
  = 0.030 376 u
∴ Binding energy = $0.030\,376 \times 931$ MeV = 28.3 MeV

(b) Binding energy per nucleon = $\dfrac{28.3}{4}$ = 7.07 MeV nuc^{-1}

Q4 (a) Power = $4\pi (1.50 \times 10^{11} \text{ m})^2 \times 1370$ W m^{-2}
  = 3.87×10^{26} W

(b) Rate of mass loss = $\dfrac{3.87 \times 10^{26} \text{ W}}{(3.00 \times 10^8 \text{ m s}^{-1})^2}$ = 4.30×10^9 kg s^{-1} = 4.3 million tonnes per second

Q5 (a) Decay constant of $^{235}_{92}$U = $\dfrac{\ln 2}{7.1 \times 10^8}$ = 9.76×10^{-10} year^{-1} = 3.09×10^{-17} s^{-1}.

No. of atoms of $^{235}_{92}$U = $\dfrac{1}{0.235} \times 6.02 \times 10^{23}$ = 2.56×10^{24}

∴ Activity = $N\lambda$ = 7.92×10^7 Bq

(b) (i) Energy is released by each decay. Nearly all the alpha particles are absorbed in the lump. This increases the vibrational energy of the $^{235}_{92}$U atoms, increasing the temperature.

(ii) Mass loss per decay = 235.043 930 − (231.036 304 + 4.002 604) u

 = 0.005 022 u

∴ Energy per decay = 4.68 MeV = 7.48×10^{-13} J

∴ Total power = 7.48×10^{-13} J × 7.92×10^{7} Bq

 = 5.9×10^{-5} W

This is undetectably small, so Michael is correct.

Q6 (a) $^{6}_{3}\text{Li} + ^{1}_{0}\text{n} \rightarrow ^{3}_{1}\text{H} + ^{4}_{2}\text{He}$

(b) $^{3}_{1}\text{H} \rightarrow ^{3}_{2}\text{He} + ^{0}_{-1}\text{e} + ^{0}_{0}\overline{\nu_e}$

(c) The reaction is $^{3}_{1}\text{H} + ^{2}_{1}\text{H} \rightarrow ^{4}_{2}\text{He} + ^{1}_{0}\text{n}$

Binding energy of $^{4}_{2}\text{He}$ = 4 × 7.1 MeV = 28.4 MeV

Binding energy of $^{3}_{1}\text{H}$ + $^{2}_{1}\text{H}$ = 3 × 2.8 + 2 × 1.1 = 10.6 MeV

∴ Energy released per reaction = 28.4 MeV - 10.6 MeV = 17.8 MeV

Appropriate mixture = 600 g $^{3}_{1}\text{H}$ + 400 g $^{2}_{1}\text{H}$, i.e. 200 moles of reactants

∴ Energy release = 6.02×10^{23} × 200 × 17.8 MeV × 1.60×10^{-13} J MeV^{-1}

 = 3.4×10^{14} J

Q7 (a) Mass loss = 4 × 1.007 825 − 4.002 604 − 2 × 0.000549 = 0.027 598 u

∴ Energy release = 0.027 598 u × 931 MeV u^{-1}

 = 25.694 MeV = 25.7 MeV (3 sf)

(b) Mass loss = 3 × 4.002 604 u − 12 u (exactly) = 0.007 812 u

∴ Energy release = 7.27 MeV

(c) (i) Luminosity = power emitted $\propto r^2 T^4$.

∴ $L = 10^2 \times 0.9^4$ = 65.6 × current value.

(ii) Energy release per triple α reaction= $\dfrac{7.27}{25.7}$ = 0.283 of that in the hydrogen fusion.

Three hydrogen fusions are needed for each triple alpha so only one-third of the reactions can take place.

∴ Energy released in triple alpha period = $\dfrac{7.27}{3 \times 25.7}$ = 0.094 × H fusion period.

∴ Lifetime ~ $\dfrac{0.094}{65.6} \times 9 \times 10^9$ years = 13 million years.

Q8 $^{56}_{26}\text{Fe} + 179\,^{1}_{0}\text{n} \longrightarrow ^{235}_{92}\text{U} + 66\,^{0}_{-1}\text{e} + 66\,^{0}_{0}\overline{\nu_e}$

179 neutrons are needed to balance the mass numbers: 56 + 179 = 235

66 electrons are needed to balance the atomic numbers: 26 = 92 + (−66)

Lepton number is conserved, so for each electron there is one anti-neutrino.

Q9 (a) The mass of two $^{4}_{2}\text{He}$ atoms = 8.005 208 u. This is less than the mass of a $^{8}_{4}\text{Be}$ atom. All the nuclear conservation laws are conserved so it does not involve the weak interaction.

(b) (i) Momentum is conserved so if the momenta of the two daughter $^{4}_{2}\text{He}$ nuclei must add to zero, i.e. they must be equal and opposite.

(ii) Mass loss = 8.005 305 − 8.005 208 u = 0.000 097 u

∴ KE release = 0.000 097 × 931 = 0.090 307 MeV = 1.44×10^{-14} J

∴ Energy of each nucleus = 0.72×10^{-14} J

∴ $7.2 \times 10^{-15} = \dfrac{1}{2} \times 4.0 \times 1.66 \times 10^{-27} v^2$

∴ $v = 1.5 \times 10^6$ m s^{-1}

Section 9: Magnetic fields

Q1 (a) I = current; ℓ = length of wire; θ = angle between the wire and field

(b) $[F]$ = kg m s^{-2}, $[I]$ = A and $[\ell]$ = m. sin θ has no unit.

∴ T = (kg m s^{-2}) A^{-1} m^{-1} = kg s^{-2} A^{-1}

(c) Fleming's left hand motor rule

Q2 (a)

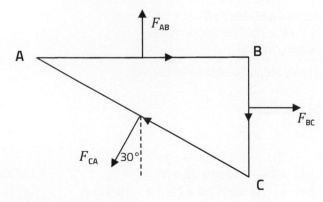

(b) (i) $\ell = 0.40 \cos 30° = 0.346$ m
∴ $F_{AB} = BI\ell = 0.030 \times 2.5 \times 0.346 = 0.026$ N

(ii) $\ell = 0.40 \sin 30° = 0.20$ m, ∴ $F_{BC} = 0.015$ N

(iii) $F_{CA} = BI\ell = 0.030 \times 2.5 \times 0.40 = 0.030$ N

(c) Vertical component of $F_{AB} = 0.030 \times 2.5 \times 0.4 \cos 30° = 0.026$ N $= F_{AC}$
∴ Resultant vertical component = 0
Horizontal component of $F_{AB} = 0.030 \times 2.5 \times 0.4 \sin 30° = 0.015$ N $= F_{BC}$
∴ Resultant vertical component = 0

Q3 (a) Motor effect force: $F = BI\ell \sin\theta$, where θ = angle between wire and field.
AB is parallel to the field so $\sin\theta = 0$. Hence $F_{AB} = 0$
BC $= \ell \sin\phi$ and $\theta = 90°$, ∴ $F_{BC} = BI\ell \sin\phi$
For CA, $F_{CA} = BI\ell \sin\phi = F_{BC}$
By Fleming's LH motor rule F_{BC} is out of the paper and F_{AB} is into the paper, both at right angles. So forces are equal and opposite, so resultant force = 0 and Ella is correct.

(b) The lines of action of F_{AB} and F_{BC} are not the same so there is a resultant moment (i.e. a couple) and the triangle would rotate – BC out of the paper and AB into the paper. Hence Ella is correct (again!).

Q4 (a) (i) Note that B_Q is parallel to PR,
B_R is parallel to PQ and B_P is parallel to QR

(ii) The resultant magnetic field is zero because the three fields are equal in magnitude and at 120° to one another.

The vector diagram is a closed triangle.

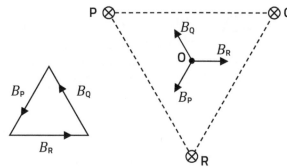

(b) (i) Field due to P at R $= \dfrac{\mu_0 I}{2\pi a} = \dfrac{4\pi \times 10^{-7} \times 7.5}{2\pi \times 0.060} = 2.5 \times 10^{-5}$ T.
∴ Force on 2.0 m of R $= BI\ell \sin\theta = 2.5 \times 10^{-5} \times 7.5 \times \sin 90° = 3.75 \times 10^{-4}$ N
Direction is towards P

(ii) Force due to Q $= 3.75 \times 10^{-4}$ N towards R.
∴ Resultant force $= 2 \times 3.75 \times 10^{-4} \cos 30°$ N $= 6.5 \times 10^{-4}$ N vertically upwards.

Q5 (a) (i) The field is uniform and strongest in the central region of the solenoid and about half this strength at the ends of the solenoid.

(ii) Insert a Hall probe into the solenoid, orientated at right angles to the axis. Keeping the current constant, measure the Hall voltage (which is proportional to the magnetic field) at a series of positions along the length of the solenoid.

(b) B – right hand grip rule or corkscrew rule.

(c) $B = \mu_0 nI = \dfrac{4\pi \times 10^{-7} \times 300 \times 4.0}{0.60} = 2.5 \times 10^{-3}$ T

(d) By inserting an iron (or other ferromagnetic) core.

Q6 (a) By conservation of energy, $eV = \frac{1}{2}mv^2$,

$$\therefore V = \frac{mv^2}{2e} = \frac{9.11 \times 10^{-31} \times (3.0 \times 10^7)^2}{2 \times 1.6 \times 10^{-19}} = 2560 \text{ V}$$

(b) $\dfrac{mv^2}{r} = Bev, \therefore B = \dfrac{mv}{er} = \dfrac{9.11 \times 10^{-31} \times 3.0 \times 10^7}{1.6 \times 10^{-19} \times 0.040} = 4.27 \times 10^{-3}$ T

Field at right angles into the diagram.

(c) If the forces balance, $Ee = Bev$, $\therefore E = 4.27 \times 10^{-3} \times 3.0 \times 10^7$ V m^{-1} = 128 kV m^{-1}
Direction = vertically downwards in the diagram.

Q7 (a) Force experienced = $Bqv \sin \theta$, at right angles to that of the velocity and field, in the direction given by the left-hand motor rule, where θ is the angle between the velocity and the field (here 90°) and v is the magnitude of the velocity.
As this force is always at right angles to the direction of motion, this provides the centripetal force.

Hence $\dfrac{mv^2}{r} = Bqv$, so $\dfrac{v}{r} = \dfrac{Bq}{m}$. But $\dfrac{v}{r} = \omega = \dfrac{2\pi}{T}$

$\therefore \dfrac{2\pi}{T} = \dfrac{Bq}{m}$ and hence $T = \dfrac{2\pi m}{Bq}$

(b) Frequency = $\dfrac{1}{\text{orbital period}} = \dfrac{0.30 \times 1.60 \times 10^{-19}}{2\pi \times 1.67 \times 10^{-27}} = 4.6 \times 10^6$ Hz

(c) The path of the protons [or other particles] is constant; the magnetic field is increased as the energy / momentum of the protons increases.
The synchrotron is used for producing very high energy (relativistic) particles, which themselves are used for probing the structure of subatomic particles.

Q8 (a) Total accelerating pd = 4×120 kV = 4.8×10^5 V
$\therefore$ Increase in kinetic energy = 4.8×10^5 eV = 7.68×10^{-14} J

$\frac{1}{2} \times 1.67 \times 10^{-27} \left(v^2 - (4.0 \times 10^6)^2\right) = 7.68 \times 10^{-14}$

$\therefore v = 1.04 \times 10^7$ m s^{-1}

(b) (i) Each time a proton emerges from a tube, the next tube must be at a lower potential (so the proton can be accelerated). This can only happen if the pd reverses polarity each time the proton is within a tube.

(ii) As the proton gains energy it speeds up, so it travels further in the time taken for the polarity to reverse.

(c) It extends over a very long distance.

Section 10: Electromagnetic induction

Q1 (a) (i) $\Phi = \pi (0.04 \text{ m})^2 \times 0.050$ T = 2.5×10^{-4} Wb

(ii) $|\varepsilon| = \dfrac{\Delta \Phi}{\Delta t} = \dfrac{2.5 \times 10^{-4} \text{ Wb}}{0.16 \text{ s}} = 1.56$ mV = 1.6 mV (2 sf)

(iii) The emf is clockwise. We know this because a clockwise emf produces a clockwise current in the ring, and hence (using the right hand grip rule) a magnetic field inside the loop directed into the paper, opposing – as required by Lenz's law – the decrease in the applied field.

(iv) Energy = power × time = $\dfrac{\varepsilon^2}{R} \Delta t = \dfrac{(1.56 \times 10^{-3} \text{ V})^2}{2.75 \times 10^{-3} \Omega} \times 0.16$ s = 1.4×10^{-4} J

(b) Doubling the diameter, d, of the ring will quadruple its area (as $A = \pi \dfrac{d^2}{4}$). So the initial flux will quadruple, and so will the induced emf, ε. The energy dissipated is proportional to ε^2/R. But R is proportional to the diameter, so the energy dissipated is 8 times as much.

Q2 (a) (i) $|\varepsilon| = \dfrac{\Delta\phi}{\Delta t} = \dfrac{B\Delta A}{\Delta t} = \dfrac{Blv}{\Delta t} = Blv$

$= 0.35\ \text{T} \times 0.15\ \text{m} \times 0.20\ \text{m s}^{-1} = 11\ \text{mV}$

(ii) $I = \dfrac{0.0105\ \text{V}}{0.020\ \Omega} = 0.53\ \text{A}$ (see diagram)

(iii) $F = BIl \sin\theta$

$= 0.35\ \text{T} \times 0.525\ \text{A} \times 0.15\ \text{m} \times \sin 90°$

$= 27.5\ \text{N} = 28\ \text{mN}$ (2 sf) (see diagram)

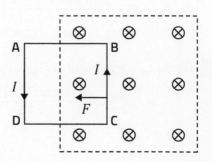

(b) Provided that the loop stays fully in the field, the flux linking it doesn't change as it moves, so there is no induced emf (Faraday's law).

(c) Left pointing arrow indicated <u>on side DA</u>.

Q3 (a) The direction of an emf induced by a change of flux is such that the effects of any current it produces oppose the change in flux.

(b) As we push the square loop into the field, there is an induced current, and energy is dissipated by resistive heating. But, as Lenz's law shows, the motor effect force on the loop is to the left, and whatever is pushing the loop has to expend energy doing work against this force. So energy is conserved.

(c) (i) Power dissipated, $\dfrac{\varepsilon^2}{R} = \dfrac{(0.0105\ \text{V})^2}{0.020\ \Omega} = 5.5\ \text{mW}$, using the emf calculated in 2(a)(i).

(ii) Work done per second $= Fv = 0.0275\ \text{N} \times 0.20\ \text{m s}^{-1} = 5.5\ \text{mW}$ using the force calculated in 2(a)(iii).

Q4 (a) (i) $\Phi = BA\cos\theta = 48 \times 10^{-6}\ \text{T} \times (0.12\ \text{m})^2 \times \cos 30° = 6.0 \times 10^{-7}\ \text{Wb}$

(ii) $N\Phi = 150 \times 6.0 \times 10^{-7}\ \text{Wb} = 9.0 \times 10^{-5}\ \text{Wb (turn)}$

(b) $|\varepsilon| = \dfrac{\Delta\Phi}{\Delta t} = \dfrac{(9.0 \times 10^{-5} - 0)\ \text{Wb-turn}}{1.2\ \text{s}} = 7.5 \times 10^{-5}\ \text{V}$

Q5 (a) When the magnetic flux linking a circuit changes an emf is induced in the circuit. The emf is proportional to the rate of change of flux linkage.

(b) (i) The flux linkage is proportional to $\cos\theta$. Therefore the rate of change of flux linkage, and hence (by Faraday's law), the emf, is proportional to $\sin\theta$. For example, when $\theta = \frac{\pi}{2}$ the flux linkage is zero, but its rate of change is a maximum, and so is the emf. The emf is also proportional to the area, PQ × QR, of the coil, the number of turns and the field strength (because these, as well as the angle, determine the flux linkage). By Faraday's law, the emf is also proportional to the rate at which the coil's angle changes, that is its angular velocity. The current is therefore proportional to all these factors, but is also inversely proportional to the sum of the resistances of the coil and resistor.

(ii) Note: from the information given, either graph is possible.

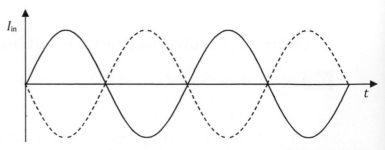

Q6 (a) As the magnet drops, the flux linked with the adjacent encircling wall of the tube changes, so an emf is induced in it, and a current, since the wall is conducting. The circular current makes a magnetic field inside the tube, roughly parallel to the coil axis. The magnet experiences a force from this field. According to Lenz's law, the force opposes the motion of the magnet.

(b) Because of the cut there can no longer be circular currents, so no magnetic field due to the induced emf and no magnetic force opposing the magnet's fall.

(c) Nabila is correct. The insulating glue will not interrupt the circular paths of currents due to the induced emfs, and there will still be a force opposing the magnet's motion, as explained in (a).

Q7 (a) Area swept out in time $\Delta t = l\, v\Delta t$ therefore in time Δt change in flux linking circuit $= \Delta\Phi = Bl\,v\Delta t$

$$\therefore \mathcal{E} = \frac{\Delta\Phi}{\Delta t} = \frac{Blv\Delta t}{\Delta t} = Blv$$

(b) (i) Initial emf $= 0.25\text{ T} \times 0.30\text{ m} \times 0.50\text{ m s}^{-1}$
$= 0.0375$ (37.5 mV) ~ 40 mV

(ii) emf $\propto l$, so increases linearly up to 75 mV. Hence (see graph):

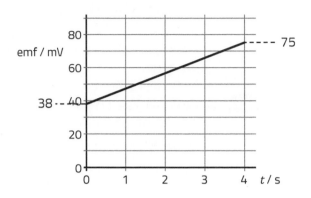

(iii) length at 3 s = 0.525 m
$\therefore$ emf = 66 mV

$$\therefore I = \frac{\mathcal{E}}{R} = \frac{66\text{ mV}}{1.50\ \Omega} = 44\text{ mA}$$

Option A: Alternating currents

Q1 (a) (i) 0.0000 s [or 0.0250 s, 0.0500 s, 0.0750 s.....]

(ii) $N\Phi = NBA = 50\text{ turn} \times 0.150\text{ T} \times (0.040\text{ m})^2$
$= 0.0120$ Wb turn

(b) (i) 0.0125 s [0.0375 s, 0.0625 s, 0.0875 s....]

(ii) $\mathcal{E}_{max} = BAN\omega = 0.0120 \times 40.0\pi = 1.51$ V [3 sf]

(c) (i) Time interval = 0.0020 s. Angle turned $= \dfrac{0.0020}{0.050} \times 360° = 14.4°$

$\therefore$ from $-7.2°$ to $7.2°$
Linkage at 0.0115 s = 0.012 sin $(-7.2°)$ = -1.50×10^{-3} Wb turn
Linkage at 0.0135 s = 1.50×10^{-3} Wb turn
$\therefore \Delta(N\Phi) = 3.00 \times 10^{-3}$ Wb turn

(ii) $\langle \mathcal{E} \rangle = \dfrac{\Delta(N\Phi)}{t} = \dfrac{3.00 \times 10^{-3}}{0.0020} = 1.5$ V

(iii) Almost the same, less than 1% difference. The rate of change of flux linkage is almost constant for small angles so to be expected.

Q2 (a) (i) $P = \dfrac{V_{rms}^{2}}{R}, \therefore V_{rms} = \sqrt{0.30 \times 5.6} = 1.30$ V

(ii) pd across the internal resistance $= \dfrac{2.4}{5.6} \times 1.30\text{ V} = 0.56$ V
$\therefore$ emf = 1.30 V + 0.56 V = 1.86 V

(b) $\mathcal{E}_{rms} = \dfrac{BAN\omega}{\sqrt{2}} = \sqrt{2}\pi BANf$

$$\therefore f = \frac{1.86}{\sqrt{2}\,\pi \times 0.30 \times (0.05)^2 \times 120}$$

$= 4.7$ Hz

Q3 (a) $V = \sqrt{V_R^2 + (V_C - V_L)^2}$

$= \sqrt{20^2 + (25-15)^2}$

$= 22.4$ V (3 sf)

(b) The reactance of the capacitor is more than that of the inductor. At resonance the two must be equal.

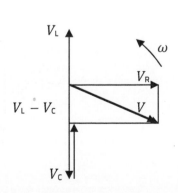

The reactance of the inductor goes up with frequency (down for the capacitor) so the resonance frequency is more than 500 Hz and Ciaran is correct.

Q4 (a) Period = 4 × 2.0 ms = 8.0 ms ∴ Frequency = 125 Hz

(b) (i) Assuming uncertainty in reading y-axis = 0.1 div
Peak pd = (1.8 ± 0.2) div × 200 mV / div = 360 ± 40 mV
∴ rms pd = 250 ± 30 mV

(ii) Using 100 mV would double the amplitude of the trace which would halve the uncertainty because the uncertainty would be 0.1 in 3.6 rather than 0.1 in 1.8. So 100 mV / div would be better.

Q5 (a) Peak pds: V_R = 1.8 V; V_C = 3.0 V
∴ rms pds: V_R = 1.3 V; V_C = 2.1 V (2 sf)

(b) Power is only dissipated in the resistor.
$$\langle P\rangle = \frac{I_0^2 R}{2} = \frac{(0.15\,A)^2 \times 12\,\Omega}{2} = 0.14\ \text{W (2 sf)}$$

Q6 (a) [Peak values]: $V = I \times \dfrac{1}{2\pi fC}$ and $f = \dfrac{1}{T}$, so
$$I = \frac{2\pi CV}{T} = \frac{2\pi \times 0.60 \times 10^{-6} \times 10}{0.020}\ \text{A} = 1.9\ \text{mA}$$

(b)
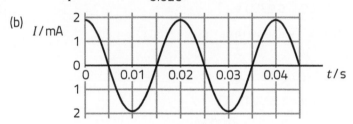

(c) From the definition of capacitance, $Q = CV$. The charging current, I, is the rate of change of Q, so is proportional to the rate of change of pd, V. Hence the maxima of the current occur at the times at which the rate of change of pd is greatest, i.e. 0 s, 0.02 s, and 0.04 s. [**Or**: the discharging current is greatest when the rate of change of pd is maximum negative, i.e. at 0.01 s and 0.03 s.]

Q7 RMS output pd, $V = \dfrac{BAN\,\omega}{\sqrt{2}}$
$$= \frac{1}{\sqrt{2}}\,45 \times 10^{-3}\,\text{T} \times 20 \times 10^{-4}\,\text{m}^2 \times 240 \times \frac{1500 \times 2\pi}{60}\ \text{s}^{-1}$$
$$= 2.40\ \text{V}$$
∴ Mean power, $\langle P\rangle = \dfrac{V_{rms}^2}{R} = \dfrac{(2.40\,\text{V})^2}{120\,\Omega} = 0.048\ \text{W}$

Q8 (a) Both have the unit ohm or both equal to $\dfrac{V_{rms}}{I_{rms}}$.

(b) Reactance (X) is frequency-dependent; resistance (R) is not
or $R = \dfrac{V(t)}{I(t)}$ but usually $X \neq \dfrac{V(t)}{I(t)}$

Q9 (a)
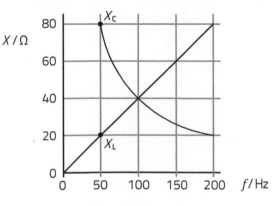

(b) (i) $X_C = \dfrac{1}{2\pi fC}$, so $C = \dfrac{1}{2\pi fX_C} = \dfrac{1}{2\pi \times 50 \times 80} = 4.0 \times 10^{-5}$ F

(ii) $X_L = 2\pi fL$, so $= L = \dfrac{X_L}{2\pi f} = \dfrac{20}{2\pi \times 50} = 0.064$ H

Q10 (a) Graph **A** is the variation of the reactance, X_L, of the inductor with frequency.

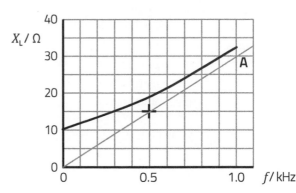

(b) At $f = 0$, $Z = 10\ \Omega$.
At $f = 0.5$ kHz, $Z = \sqrt{10^2 + 15^2} = 18.0\ \Omega$
At $f = 1.0$ kHz, $Z = \sqrt{10^2 + 30^2} = 31.6\ \Omega$

Q11 (a) (i)

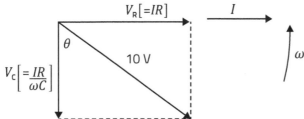

(ii) Applying Pythagoras' theorem: $10\ \text{V} = \sqrt{V_R^2 + V_C^2} = I\sqrt{R^2 + X_C^2}$

$X_C = \dfrac{1}{2\pi \times 750 \times 0.47 \times 10^{-6}}\ \Omega = 450\ \Omega$

$\therefore I = \dfrac{10.0}{\sqrt{330^2 + 450^2}}\ \text{A} = 0.018\ \text{A} \sim 20\ \text{mA}$

(b) (i) $V_C = IX_C = 0.018 \times 450 = 8.1$ V

(ii) Angle θ in phasor diagram $= \tan^{-1}\left(\dfrac{V_R}{V_C}\right) = \tan^{-1}\left(\dfrac{R}{X_C}\right) = \tan^{-1}\left(\dfrac{330}{450}\right) = 36°$

(iii) This is not true: $V_R = \sqrt{10.0^2 - V_C^2} = 5.9$ V which is not the same as $(10.0 - 8.1)$ V.

(c) (i) Only the resistor dissipates energy.
Power dissipated $= I^2 R = 0.105$ W

$\therefore$ Energy over one cycle $= \dfrac{0.105\ \text{W}}{750\ \text{Hz}} = 1.4 \times 10^{-4}$ J (2 sf)

(ii) Peak pd across capacitor = 11.5 V
$\therefore$ Maximum energy stored $= \frac{1}{2}CV^2 = 3.1 \times 10^{-5}$ J
$\therefore$ Mean energy = 1.5×10^{-5} J

Q12 (a) Set the signal generator to 100 Hz. Take simultaneous readings of the current and pd using the ammeter and voltmeter respectively. Divide the pd by the current to give the impedance.

(b)

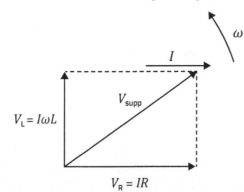

From the phasor diagram: $V_{\text{supp}} = \sqrt{V_L^2 + V_R^2} = I\sqrt{X_L^2 + R^2}$
By definition $V_{\text{supp}} = IZ$ $\therefore Z^2 = X_L^2 + R^2 = (2\pi fL)^2 + R^2$

(c) (i) Intercept = 25 Ω^2

$\therefore R = 5.0\ \Omega$

(ii) Gradient = $\dfrac{231 - 25}{10 \times 10^5}$ = $2.06 \times 10^{-4}\ \Omega^2\ s^2$

Gradient = $4\pi^2 L^2$, $\therefore L^2 = 5.22 \times 10^{-6}\ H^2$

$\therefore L = 2.3\ mH$

Q13 (a) At resonance, pd across $R = 5.00$ V

$\therefore R = \dfrac{5.00}{1.00}\ \Omega = 5.00\ \Omega$

Resonance frequency $f_{res} = \dfrac{1}{2\pi\sqrt{LC}}$. $\therefore L = \dfrac{1}{4\pi^2 f_{res}^2 C}$ = 2.25×10^{-3} H

(b) (i) $V_C = \dfrac{I}{2\pi fC} = \dfrac{1.00}{2\pi \times 10.6\ \times 10^3 \times 0.100 \times 10^{-6}}$ = 150 V

(ii) $Q = \dfrac{150}{5} = 30$

Q14 $L = 4\pi \times 10^{-7}\ Hm^{-1} \times \dfrac{25^2 \times \pi \times (3.0 \times 10^{-3}\ m)^2}{15 \times 10^{-3}\ m}$ = 1.48×10^{-6} H

$f_{res} = \dfrac{1}{2\pi\sqrt{LC}}$, so $= C = \dfrac{1}{4\pi^2 f_{res}^2 L} = \dfrac{1}{4\pi^2 + (1.6 + 10^6\ Hz)^2 + 1.48 + 10^{-6}H}$ = 6.7 nF

Q15 (a)

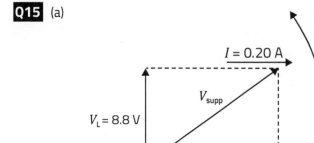

$V_R = 0.20$ A $\times$ 33 Ω = 6.6 V

$X_L = 2\pi \times 100$ Hz $\times 0.070$ H = 44.0 Ω

$\therefore V_L = 0.20$ A $\times$ 44 W = 8.8 V

$\therefore V_{supp} = \sqrt{6.6^2 + 8.8^2}$ = 11 V

(b) (i) $f_{res} = \dfrac{1}{2\pi\sqrt{LC}}$

$\therefore C = \dfrac{1}{4\pi^2 f_{res}^2 L} = \dfrac{1}{4\pi^2 \times (100\ Hz)^2 \times 0.070H}$

$= 3.6 \times 10^{-5}$ F = 36 µF

(ii) $I = \dfrac{V}{R} = 0.30$ A

(c) He will be correct as far as the peak current is concerned because both the pd and resistance are doubled. However the resonance curve will be less sharp (the Q factor will be less) because the ratio of the pds across the inductor (or capacitor) and resistor will be less.

Option B: Medical physics

Q1 X-ray attenuation in tissues increases with increasing density, so bones (high density) cause more attenuation than soft tissues, making good contrast images. X-rays have very low wavelength so diffraction is negligible, so produce sharp images. [They can also be detected by CCD devices, photographic film or MOSFET devices.]

Q2 (a) The electrons are accelerated to a high energy by the pd. These high energy (70 keV) electrons decelerate rapidly in the tungsten target. Charged particles which accelerate produce radiation [by the process of *bremsstrahlung*] giving the continuous spectrum.

(b) The electrons sometimes knock out an inner electron from a tungsten atom. Higher energy electrons within the atoms drop down into the vacant energy level giving out photons with energy equal to the difference in the two energy levels.

(c) [A vacuum is needed] to allow the electrons to pass down the tube without colliding with air molecules and losing energy.

(d) $\frac{1}{2}mv^2 = eV$, so $v = \sqrt{\dfrac{2eV}{m}} = \sqrt{\dfrac{2 \times 1.60 \times 10^{-19}\,C \times 70 \times 10^3\,V}{9.11 \times 10^{-31}\,kg}} = 1.6 \times 10^8\ m\ s^{-1}$

This calculated speed is too close to the speed of light (in a vacuum) for the above equations to be valid.

(e) $eV = \dfrac{hc}{\lambda_{min}}$, so $\lambda_{min} = \dfrac{hc}{eV} = \dfrac{6.63 \times 10^{-34}\,Js \times 3.00 \times 10^8\,m\,s^1}{1.60 \times 10^{-19}\,C \times 70 \times 10^3\,V} = 1.8 \times 10^{-11}\ m$

(f) (i) Power input = 14.5 mA × 70 kV = 1015 W

∴ Efficiency = $\dfrac{5.1\ W}{1015\ W}$ = 5.0 × 10⁻³ [= 0.5%]

(ii) Heat needs to be conducted away at the rate of 1015 W - 5.1 W, i.e. about 1 kW. In the absence of cooling water the temperature of the tungsten cathode would become too high and the X-ray tube would be destroyed (the glass would melt).

(iii) Predicted % efficiency = 70 × 74 × 10⁻⁴ = 0.52 (2 sf)
So this is a very good approximation.

Q3

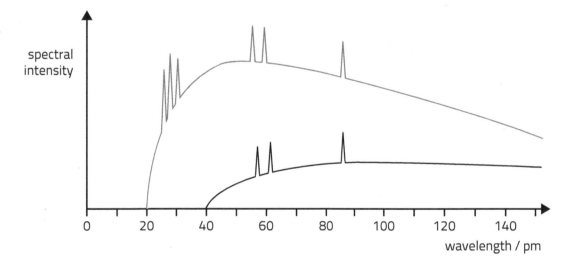

(horizontal axis) wavelength / pm

Q4 (a) (i) In the piezoelectric effect, distorting a crystal produces a pd. If a pd is applied to a piezoelectric crystal, it produces a distortion – the reverse piezoelectric effect.

(ii) A short-duration high frequency (MHz) voltage applied to the crystal in the transducer causes rapid vibrations (of the same frequency) producing the ultrasound wave.

(b) A-scans are amplitude scans. The returning wave is displayed on a CRO and the time delay indicates the depth of the structure.
B-scans are brightness scans. The returning waves build up an image of the structures on a screen. [This is achieved using a large array of ultrasound detectors. 2D images are built up as the emitter and detectors are rotated/scanned.]

(c) (i)

Body tissue	Density / kg m⁻³	Speed of sound / m s⁻¹	Acoustic impedance / kg m⁻² s⁻¹
Soft tissue	1070	1590	1.70 × 10⁶
Bone	1650	4080	6.73 × 10⁶

(ii) $R = \dfrac{(6.73 - 1.70)^2}{(6.73 + 1.70)^2}$ =0.356 = 36% (2 sf)

[Note – the factors of 10⁶ cancel so there is no need to include them in the calculation.]

(iii) There are four transitions between soft tissue and bone: skin/bone; bone/brain; brain/ bone; bone/skin. The fraction transmitted at each boundary = 0.644, so for four boundaries the transmission is $(0.644)^4$ = 0.172 = 17%. Hence the signal is not 13% of the baby's signal. So, the final figure is approximately correct, but this is a fluke as it does not arise from $1.00 - 0.36^2$.

Q5 (a) μ is defined by: $I = I_o e^{-\mu x}$ where I is the intensity.

If $I = \frac{1}{2}I_o$, then $e^{-\mu x_{1/2}} = \frac{1}{2}$

Taking natural logs $\rightarrow -\mu x_{1/2} = \ln\frac{1}{2} = -\ln 2$

$\therefore \mu x_{1/2} = \ln 2$

(b) $\mu = \dfrac{\ln 2}{3.7 \text{ cm}} = 0.1873 \text{ cm}^{-1}$

$\therefore \dfrac{I}{I_0} = e^{-0.1873 \times 5.0} = 0.39$

$\therefore$ Fractional reduction = 1.00 − 0.39 = 0.61 = 61%

Q6 (a) The scanner head produces pulses of ultrasound which penetrate the body and are reflected from the foetus. The reflected waves are detected by the scanner head, which gives out voltage signals to an analyser which builds up an image as the scanner is moved across the abdomen. The gel is required so that the reflection coefficient between the head and the abdomen is low − otherwise the air gap would produce large reflections and very little signal would get through to the foetus.

(b) (i) The ultrasound rays are reflected from the red blood cells back to the scanner head. If the blood cells are moving towards the scanner, the waves are Doppler shifted to a higher frequency: the blood cells 'see' waves of a higher frequency which they reflect; as the cells are a moving source the waves received by the scanner are further Doppler shifted. The frequency change is proportional to the speed of the blood cells.

(ii) $\dfrac{\Delta f}{f_0} = -\dfrac{2 \times 1.058 \text{ m s}^{-1}}{1580 \text{ m s}^{-1}}\cos 5° = 1.334 \times 10^{-3}$

$\therefore \Delta f = 1.334 \times 10^{-3} \times 5.50 \text{ MHz} = 7.34 \text{ kHz (3 sf)}$

Q7 (a) The anti-scatter grid is to cut out X-rays which are diverted at an angle by the material of the body − which interfere with the image.

The scintillator screen flashes at a point where an X-ray photon hits to produce the image.

(b) They are designed so that each X-ray photon produces a large number of visible photons, which reduces the dose of X-rays required. The screen is placed in a dark container so that even a very faint image will be picked up by the camera − also cutting down the X-ray dose required.

Q8 (a) In therapy, the photons are required to deliver a large energy to the target cells. Higher energy X-ray photons have a greater penetration. This means that the intensity of the beam drops more slowly, giving a more uniform dosage to the desired area.

(b) In therapy, the X-rays are required to cause damage to the target (tumour) cells. Higher intensity beams deliver a greater number of photons per second, increasing the damage to the target cells.

Q9 (a) **Photon explanation:**
The hydrogen nuclei have two spins (up and down in the magnetic field) with an energy difference, and more nuclei in the lower energy level. The radio photons have energy equal to this difference and promote a higher number of nuclei into the upper energy level. When the radio beam is turned off, the subsequent release of energy as the nuclei return to the lower energy level is detected.
Classical explanation:
The spinning hydrogen nuclei precess at a frequency proportional to the magnetic field strength. If radio waves with this precession frequency are incident they are strongly absorbed and re-emitted when the external radio waves are turned off.

(b) $f = 62.4 \, B$
$f_1 = 62.4 \times 1.53 = 95.5 \text{ MHz}$; $f_2 = 62.4 \times 1.92 \text{ MHz} = 120 \text{ MHz}$

(c) **MRI scans** produce high resolution (around 1 mm) 3D images in which the different tissues in the joint are distinguished. They require the use of very expensive equipment and can cause claustrophobia. They carry no known risk but cannot be given to people with metal implants, e.g. heart pacemakers.

Traditional X-rays only image bones clearly, with excellent image resolution (~0.1 mm) and are 2D. So the soft tissues (cartilage and ligaments) in the joint are not imaged well. They also have a small risk because of the ionising radiation involved. [3D and soft tissue images can be produced by CT X-ray scans with the dangers of higher radiation dose. Image resolution is around 0.5 mm.]

Ultrasound B-scans are cheap and benign but have poor resolution (usually around 2-5 mm) and can image the soft tissues and the surface of bones (so are good for checking cartilage and ligament).

Q10 (a) Additional distance from **X** to detector **B** = 3.00×10^8 m s^{-1} × 237 ps = 7.11 cm
∴ **X** is 3.6 cm closer to **A** than the halfway point.

(b) A positron emitter is attached to glucose molecules where it is preferentially absorbed by actively dividing cells, such as tumour cells. The emitted positrons have short range and mutually annihilate with an electron in the body to produce 2 γ photons which emerge in opposite directions. The time difference between the detection of these photons enables the position of the emission site to be determined. A (3D) image is gradually built up showing the location of hot-spots where the annihilations take place.

Q11 The X-ray-emitting head and the detector rotate about the body and gradually move along the axis of the body. In this way the scanner produces images of the body in slices which can be combined in a computer to produce 3D images. Traditional X-rays don't distinguish soft tissues well but contrast agents can be injected into blood vessels or swallowed into the alimentary canal to improve the visibility of different soft tissues.

Q12 (a) (i) Effective dose = equivalent dose × tissue weighting factor
Effective dose (contribution for the liver) = 550 mSv × 0.04 = 22 mSv

(ii) Equivalent dose = absorbed dose × radiation weighting factor
Weighting factor for these neutrons = 20

Absorbed dose = $\dfrac{550}{20}$ = 27.5 mGy

Absorbed dose = $\dfrac{\text{energy of absorbed radiation}}{\text{mass}}$

∴ Energy of absorbed radiation = absorbed dose × mass
= 27.5 mGy × 94 kg = 2.6 J

(b) The radiation weighting factor of a radiation is 20, which is the highest, reflecting the fact that it is strongly ionising and hence it deposits its energy in a very short distance. The ingested material will come into close contact with the oesophagus, the stomach and the colon (even assuming none is absorbed into the bloodstream) which presents a high risk of cancers to the lining of these organs. Together these organs make up about 30% of the sensitivity of the body to radiation.

Q13 (a) The collimator is there to restrict the γ rays to those in the vertical direction, to allow an image to be made (because γ rays cannot be focused).

The scintillator produces a flash with many visible photons when one γ photon hits it allowing an image to be constructed. The photomultiplier is very sensitive and ensures that each of these flashes is detected to allow a low dose of ^{99m}Tc to be used.

(b) It needs to be non-toxic, have a short half-life and to produce radiation which is not absorbed in the body, which can hence emerge to be detected, i.e. a gamma emitter.

Option C: The physics of sports

Q1 Football is a contact sport in which players receive forces from the side. The shorter the height, h, (see diagram) the smaller the moment of F and the wider the base w, the bigger the opposing moment to being toppled. Hence the greater the stability.

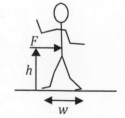

Q2 Clockwise moment of leg + foot = $118 \times 43 + 11 \times 91$ N cm

$$= 6075 \text{ N cm}$$

∴ By, principle of moments, $F \times 4.5$ cm = 6075 N cm

∴ $F = \dfrac{6075 \text{ N cm}}{4.5 \text{ cm}} = 1350$ N

Q3 (a) (i) Total $E_k = \sum \frac{1}{2}(\Delta m) v^2$, where the Δm are the small masses that make up the cylinder.

Every part of the cylinder is moving with the same speed, $v = rw$.

∴ $E_k = \frac{1}{2}v^2 \sum \Delta m = \frac{1}{2}(r\omega)^2 \sum \Delta m$

But $\sum \Delta m$ = total mass, m. ∴ $E_k = \frac{1}{2} mr^2\omega^2$

(ii) For a rolling cylinder moving with speed v, the angular velocity, $\omega = \dfrac{v}{r}$, so that the point of contact is stationary.

kinetic energy $E_k = \frac{1}{2} mv^2 + \frac{1}{2} I\omega^2$

But $I = \frac{1}{2} mr^2$ ∴ $\frac{1}{2} I\omega^2 = \frac{1}{2} mr^2\omega^2 = \frac{1}{2} m(r\omega)^2 = \frac{1}{2} mv^2$

Hence the rotational and linear KEs are equal, i.e. they contribute half each.

(iii) Loss of gravitational potential energy = $mg\Delta h$

Total KE = 2×0.45 J = 0.90 J

∴ By conservation of energy, $m = \dfrac{0.90 \text{ J}}{9.81 \text{ N kg}^{-1} \times 0.30 \text{m}} = 0.31$ kg (2 sf)

(b) For a particular mass and radius, the moment of inertia of the snooker ball is less than that of the cylinder, so the rotational kinetic energy of a rolling snooker ball is less than its translational kinetic energy. So for a particular potential energy loss the translational kinetic energy of the snooker ball is greater than that of the cylinder, so it has a greater speed and hence it arrives first.

Q4 (a) Linear acceleration, $a = \dfrac{[48 \text{ m s}^{-1} - (-32 \text{m s}^{-1})]}{0.0068 \text{ s}} = 11\,800$ m s^{-2} Due west

$\Delta\omega = \dfrac{2550 - 1200}{60}$ rad s^{-1} = 22.5 rad s^{-1}

∴ Angular acceleration, $a = \dfrac{22.5 \text{ rad s}^{-1}}{0.0068 \text{ s}} = 3310$ rad s^{-2}

(b) Mean force, $\langle F \rangle = ma = 0.0582$ kg $\times 11\,800$ m s^{-2} = 687 N

Mean torque, $\langle \tau \rangle = I\alpha = \frac{2}{3} \times 0.0582$ kg $\times \left(\dfrac{66.9 \times 10^{-3} \text{ m}}{2}\right)^2 \times 3310$ rad s^{-2} = 0.144 N m

(c) Rotational kinetic energy,

$E_{k\,rot} = \frac{1}{2} I\omega^2 = \frac{1}{2} \times \frac{2}{3} \times 0.0582$ kg $\times \left(\dfrac{66.9 \times 10^{-3} \text{ m}}{2}\right)^2 \times \left(\dfrac{2550 \text{ rad}}{60 \text{ s}}\right)^2 = 0.039$ J

Translational kinetic energy,

$E_{k\,trans} = \frac{1}{2} mv^2 = \frac{1}{2} \times 0.0582$ kg $\times (48 \text{ m s}^{-1})^2 = 67$ J

So Charles is correct by a large margin!

(d) Consider the vertical motion with upwards positive

To calculate the time to hit the ground: $y = u_y t - \frac{1}{2} gt^2$, with $y = -0.95$ m

∴ $4.905t^2 - 48t \sin 6.5° - 0.95 = 0$, ∴ $t = \dfrac{5.43 \pm \sqrt{29.5 + 4 \times 4.905 \times 0.95}}{9.81} = 1.26$ s

[ignoring the negative root]. ∴ Range = $(48 \cos 6.5° \times 1.26)$ m = 60 m (2 sf)

So she has hit it too far.

(e) (i) $F_D = \frac{1}{2}\rho v^2 A C_D$

$= \frac{1}{2} \times 1.25$ kg m$^{-3} \times (48 \text{ m s}^{-1})^2 \times \pi \left(\dfrac{66.9 \times 10^{-3} \text{ m}}{2}\right)^2 \times 0.60 = 3.0$ N (2 sf)

Acting over a distance of 10 m this would reduce the KE by 30 J which is nearly half so it cannot be ignored.

(ii) Gravitational force = mg = 0.57 N, so the 'lift' is almost 4× as great and the range will be a lot less than the calculated 60 m.

Q5 The shower spray makes the air within the shower cubicle move. This causes the pressure within the cubicle to drop according to the Bernoulli equation, $p = p_0 - \frac{1}{2}\rho v^2$. The pressure difference between the inside and outside produces a net inward force on the shower curtain.

Q6 The anticlockwise moment of the weight of the wind surfer (the person and the vessel) about the contact with the water (the turning point, **T**) is balanced by the clockwise moment of the force of the wind on the sail.

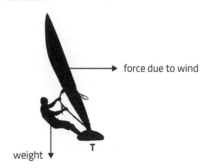

force due to wind

weight **T**

Q7 Speed of approach = 11.8 m s^{-1}
Speed of separation = (11.4 − 0.4) m s^{-1} = 11.0 m s^{-1}
∴ Coefficient of restitution, $e = \dfrac{11.0 \text{ m s}^{-1}}{11.8 \text{ m s}^{-1}} = 0.93$

Q8 In the absence of external torques, the angular momentum, L, of the gymnast is constant.
$L = I\omega$, where $I = \sum mr^2$ is her moment of inertia.
In the tuck position, the distance r from the centre to many points of the body is reduced. Hence, I is reduced and ω, the angular velocity, increases.

Q9 (a) Momentum change,
Δp = 0.058 kg × [49 m s^{-1} − (−63 m s^{-1})] m s^{-1} = 6.50 N s due south.
∴ Mean force on ball, $\langle F \rangle = \dfrac{\Delta p}{t} = \dfrac{6.5 \text{ N s}}{6.5 \times 10^{-3} \text{ s}} = 1.0 \times 10^3$ N

(b) (i) $e = \dfrac{49 \text{ m s}^{-1}}{63 \text{ m s}^{-1}} = 0.78$

(ii) $e = \sqrt{\dfrac{\text{bounce height}}{\text{drop height}}}$, ∴ $\dfrac{\text{bounce height}}{\text{drop height}} = 0.78^2 = 0.60$

Q10 (a) $\langle F \rangle = \dfrac{m\Delta v}{t} = \dfrac{0.04593 \text{ kg} \times 85 \text{ m s}^{-1}}{257 \times 10^{-6} \text{ s}} = 15\,200$ N

(b) $\Delta\omega = \dfrac{2700}{60} \times 2\pi$ rad s^{-1} = 283 rad s^{-1}

∴ $\langle\alpha\rangle = \dfrac{\Delta\omega}{t} = \dfrac{283}{257 \times 10^{-6}}$ rad s^{-2} = 1.10×10^6 rad s^{-2}

(c) Mean torque, $\langle\tau\rangle = I\alpha$
∴ $\langle F_{\text{tan}}\rangle r = \frac{2}{5}mr^2\langle\alpha\rangle$,
so $\langle F_{\text{tan}}\rangle = \dfrac{2mr}{5}\langle\alpha\rangle = \dfrac{2 \times 0.04593 \text{ kg} \times 21.34 \times 10^{-3}\text{m}}{5} \times 1.10 \times 10^6$ rad s^{-2}
= 431 N

(d)

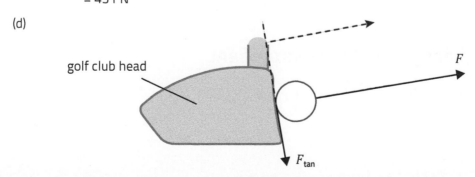

golf club head

F

F_{tan}

(e) $E_{k\,rot} = \frac{1}{2}I\omega^2 = \frac{1}{2} \times \frac{2}{5} \times 0.04593\,kg \times (21.34 \times 10^{-3}\,m)^2 \times (283\,rad\,s^{-1})^2 = 0.34\,J$

$E_{k\,trans} = \frac{1}{2}mv^2 = \frac{1}{2} \times 0.04593\,kg \times (85\,m\,s^{-1})^2 = 166\,J$

$\therefore \dfrac{E_{k\,rot}}{E_{k\,trans}} = \dfrac{0.34\,J}{166\,J} = 2.0 \times 10^{-3}$

Q11 (a) Initial vertical velocity = $18.1 \sin 21.2°\,m\,s^{-1} = 6.545\,m\,s^{-1}$

Calculate time in air: $t = \dfrac{v-u}{a} = \dfrac{6.545\,m\,s^{-1} - (-6.545\,m\,s^{-1})}{9.81\,m\,s^{-1}} = 1.334\,s$

Horizontal velocity = $18.1 \cos 21.2° = 16.88\,m\,s^{-1}$

$\therefore$ Range = 22.5 m

(b) $F_D = \frac{1}{2}\rho v^2 A C_D$

$= \frac{1}{2} \times 1.25\,kg\,m^{-3} \times (18.1m\,s^{-1})^2 \times \pi\,(0.110\,m)^2 \times 0.195 = 1.52\,N$

(c) Because of the spin, the air passing round the ball is deflected downwards as shown in the diagram, acquiring a downward momentum. So the ball exerts a downward force on the air (N2) and the air exerts an equal upward force on the ball (N3) so the flight time of the ball is greater.

(d)

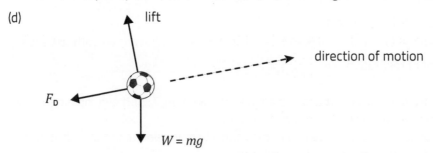

(e) The drag acts in the opposite direction to the motion of the ball through the air. In the absence of spin it produces a deceleration both horizontally and vertically. The vertical deceleration has a very small effect on the flight time because, although the mean upward velocity is less, the mean downward velocity is reduced further. The range, however, is reduced because of the horizontal deceleration producing a lower mean speed.

Backspin produces a lift force which is mainly vertical upwards, this increases the time the ball spends in the air and hence increases the horizontal range.

(f) Angular momentum is conserved if there is no externally applied torque [couple]. In this case, there is a couple applied to the ball by the air, so the principle does not apply to the ball on its own.

(g)

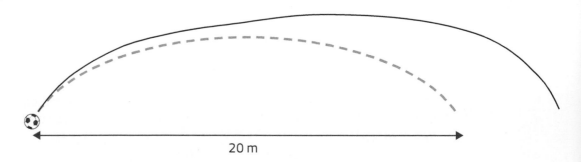

Option D: Energy and the environment

Q1 (a) (i) Cross-sectional area of asteroid = $\pi r^2 = \pi \times (150m)^2$

$\therefore$ Radiation absorption rate = $0.9 \times \pi \times (150\,m)^2 \times 1361\,W\,m^{-2}$

$= 86.6\,MW$

(ii) If the asteroid is rotating, the surface temperature will be approximately the same on both faces. If it radiates as a black body, Stefan's law is valid.

i.e. $P = A\sigma T^4$, with $A = 4\pi r^2$

$$\therefore T^4 = \frac{86.6 \times 10^6 \text{ W}}{4\pi \times (150 \text{ m})^2 \times 5.67 \times 10^{-8} \text{ W m}^{-2} \text{ K}^{-4}}$$

$\therefore T = 271$ K

Alternatively:

$0.9 \times \pi \times 150^2 \times 1361 = 4\pi \times 150^2 \times 5.67 \times 10^{-8} T^4$

leading to the same answer

(iii) $\lambda_{\text{peak}} = \dfrac{W}{T} = \dfrac{2.90 \times 10^{-3} \text{ m K}}{271 \text{ K}} = 10.7$ μm

Infra-red.

(b) (i) The radiation from the surface passes through the atmosphere. This radiation is in the thermal infra-red and is partly absorbed by greenhouse gases in the atmosphere. This absorbed energy is re-radiated, in all directions, with 50% downwards where it is absorbed by the Earth, resulting in a higher stable temperature.

(ii) Higher concentrations result in a greater fraction of the radiation from the ground being absorbed and re-radiated downwards. Thus the ground temperature rises.

Q2 (a) A body which is wholly or partially immersed in a fluid experiences an upthrust equal to the weight of fluid displaced by the body.

(b) (i) Mass of ice = 920 kg m^{-3} × 1.00 × 10^{-4} m^3

= 0.092 kg

$\therefore$ The mass of water displaced = 0.092 kg

$\therefore$ Volume of water displaced = $\dfrac{0.092 \text{ kg}}{1000 \text{ kg m}^{-3}} = 92$ cm^3

(ii) When the ice melts, the melt water has the same density as the water in the can so occupies a volume of 92 cm^3. Hence the water level stays the same. The principle is the same for a floating iceberg or ice-shelf. Hence melting sea ice hardly changes the sea level.

(c) The ocean surface is darker than ice, as is the rock exposed by retreating glaciers. Hence the fraction of the Sun's radiation absorbed by the Earth increases, raising the temperature further.

Q3 Some renewable sources, e.g wind, are intermittent. Solar energy has a reliable daily and seasonal variation but in Britain has a large intermittent element. Tidal power is renewable but almost totally dependable. It is not the intermittence which is the problem but the inability to store the generated energy, which could be overcome by using it produce a fuel, e.g. hydrogen by electrolysis of water.

Q4 (a) A fuel cell is a device in which the chemical energy in a fuel is used directly in a non-thermal process for the production of electrical energy.

(b) Fuel cells can be run from hydrogen, with oxygen from the atmosphere as the oxidising agent, so produce only water vapour as an exhaust gas. This is an advantage if the hydrogen is generated using electricity from a renewable source (or any CO_2 is sequestered). In conjunction with an electric motor, fuel cells are much more efficient than internal combustion engines.

Q5 (a) Uranium enrichment is increasing the percentage of ^{235}U in a sample of uranium from the natural level of 0.7%. It is necessary because the majority isotope, ^{238}U, is non-fissile but rather absorbs neutrons without fission. Fission reactors need about 3% of ^{235}U for normal operation.

(b) (i) The thorium nuclide must be $^{232}_{91}$Th.

$$^{232}_{91}\text{Th} + {}^{1}_{0}\text{n} \rightarrow {}^{233}_{91}\text{Th} \rightarrow {}^{233}_{92}\text{U} + {}^{0}_{-1}\text{e} + {}^{0}_{0}\overline{\nu}_e$$

(ii) A ^{238}U nucleus absorbs a neutron, producing ^{239}U. This decays in two stages by β$^-$ emission producing ^{239}Np followed by ^{239}Pu. The reactions are:

$$^{238}_{92}\text{U} + {}^{1}_{0}\text{n} \rightarrow {}^{239}_{92}\text{U}$$

$$^{239}_{92}U \rightarrow {}^{239}_{93}Np + {}^{0}_{-1}e + {}^{0}_{0}\overline{\nu}_e$$

$$^{239}_{93}Np \rightarrow {}^{239}_{94}Pu + {}^{0}_{-1}e + {}^{0}_{0}\overline{\nu}_e$$

Q6 (a)

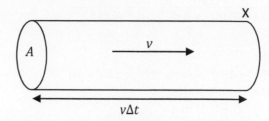

The mass of air passing the cross-section at **X** in time $\Delta t = Av\Delta t \times \rho$

∴ Kinetic energy of this air $= \frac{1}{2}(A\rho v\Delta t)v^2 = \frac{1}{2}A\rho v^3\Delta t$

∴ Dividing by Δt, the KE/second $= \frac{1}{2}A\rho v^3$

(b) Efficiency $= \dfrac{P_{OUT}}{P_{IN}} \times 100\%$

$= \dfrac{7.9 \times 10^6 \text{ W}}{\frac{1}{2}\pi(80\text{ m})^2 \times 1.25 \text{ kg m}^{-3} \times (12 \text{ m s}^{-1})^3} \times 100\%$

$= 36\%$

Q7 (a) $\dfrac{\Delta Q}{\Delta t}$ = the heat flow per unit time (i.e. the power transfer).

Unit: W

$\dfrac{\Delta \theta}{\Delta x}$ = the temperature gradient, i.e. the temperature difference per unit length:

Unit: K m^{-1} or °C^{-1} m^{-1}

(b) $\left[\dfrac{\Delta Q}{\Delta t}\right] = [A][K]\left[\dfrac{\Delta \theta}{\Delta x}\right]$,

∴ W $= \text{m}^2 [K] \text{ K m}^{-1}$

∴ $[K] = \dfrac{\text{W}}{\text{m}^2 \text{ K m}^{-1}} = \text{W m}^{-1}\text{K}^{-1}$

(c) Heat flows against the temperature gradient, i.e. from a high to a low temperature.

(d) $\Delta \theta = 25\,°C = 25$ K

$\dfrac{\Delta Q}{\Delta t} = 0.5 \text{ m}^2 \times 0.14 \text{ W m}^{-1} \text{K}^{-1} \times \left(\dfrac{25 \text{ K}}{12.0 \times 10^{-3} \text{ m}}\right)$

$= 146 \text{ W} = 8750 \text{ J / minute}$

Q8 (a) The intensity of the solar radiation at the ground is much less than 1361 W m^{-2} because of absorption and scattering by the atmosphere.

(b) At 15 V, in units of A and W m^{-2}: $\dfrac{9.2}{400} = 0.023$; $\dfrac{14.2}{600} = 0.024$; $\dfrac{19.2}{800} = 0.024$; $\dfrac{23.8}{1000} = 0.024$.

Hence proportional, within the tolerance of graphing.

(c) $P = VI$. At **X** $V = 0$, ∴ $P = 0$

At **Y** $I = 0$, ∴ $P = 0$.

(d)

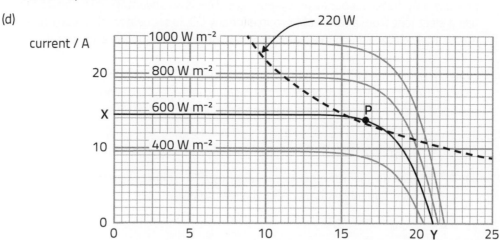

The broken line on the grid is for 220 W. The point **P** (15.5 V, 11.9 A) on the 600 W m^{-2} characteristic is slightly above the 220 W line, so the manufacturer's claim is correct.

The power at **P** is 16.5 V × 11.9 A = 229 W.

Q9 (a) (i) The kinetic energy of the water issuing per second = $\frac{1}{2}A\rho v^3$

But $v = \dfrac{\text{volume per second}}{\text{cross-sectional area}} = \dfrac{2.75 \text{ m}^3 \text{ s}^{-1}}{\pi \times (0.50 \text{ m})^2} = 3.50 \text{ m s}^{-1}$

∴ KE gain per second = $\frac{1}{2}\pi(0.5 \text{ m})^2 \times 1000 \text{ kg m}^{-3} \times (3.50 \text{ m s}^{-1})^3$

= 16 850 W

(ii) Mass flow = 2750 kg s^{-1}

∴ Loss in GPE per second = 2750 kg s^{-1} × 9.81 N kg^{-1} × 6.0 m

= 162 000 W

(b) Available power = 162 kW – 17 kW

= 145 kW

∴ Power output = 80% × 145 kW = 116 kW

(c) KE gained per second = 16.85 kW × (1.10)^3 = 22.43 kW

GPE loss per second = 162 kW × 1.10 = 178 kW

∴ Power output = 80% × (178 – 22) kW = 125 kW

110% of 116 kW = 128 kW,

∴ output a bit less than 10% more

Original efficiency = $\dfrac{116 \text{ kW}}{162 \text{ kW}} \times 100 = 72\%$ (2 sf)

With 10% more flow: Efficiency = $\dfrac{125 \text{ kW}}{178 \text{ kW}} \times 100\% = 70\%$ (2 sf)

∴ Overall efficiency is lower but less than 10% lower.

Q10 The first stage is a weak interaction, shown by the emission of the neutrino, which is therefore much less likely to occur in any suitable collision than the second stage which is [a strong interaction followed by] an electromagnetic interaction.

Q11 (a) The rate of heat loss through each m^2 of the wall is 0.18 W for every °C difference in temperature between the inside and the outside.

(b) Rate of heat loss through wall = (7.0 × 2.2 – 3.0) m^2 × 0.18 W m^{-2} K^{-1}

= 2.23 W K^{-1}

Rate of heat loss through double-glazed units = 4.5 W K^{-1}

Rate of heat loss through triple-glazed units = 2.4 W K^{-1}

∴ % reduction = $\dfrac{4.5 - 2.4}{4.5 + 2.2} \times 100 = 31\%$

(c) The mean outdoor temperature is much lower in Norway, so the heat loss is much greater. Hence the payback time of installing triple glazing is much shorter.

Q12 (a) Rate of heat loss = 8.0 m^2 × 0.62 W m^{-1} K^{-1} × $\dfrac{0.35 \text{ K}}{0.10 \text{ m}}$ = 17 W

(b) Temperature difference across insulation = $\dfrac{0.62}{0.039} \times 0.35 \text{ °C} = 5.56 \text{ °C}$

∴ Total temperature difference = 0.35 + 5.56 + 0.35 = 6.3 °C (2 sf)

(c) (i) 17 W = 8.0 m^2 × (22 – 8)°C × U

∴ $U = \dfrac{17 \text{ W}}{8.0 \text{ m}^2 \times 16°\text{C}} = 0.13 \text{ W m}^{-2} \text{ K}^{-1}$

(ii)

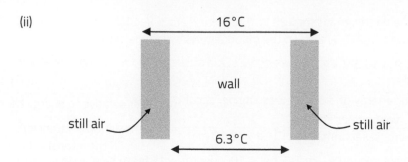

On either side of the wall there is a layer of still air which has a very low thermal conductivity and hence a large temperature difference across it.

(iii) The layer of still air on the outer face of the wall would be blown away so the temperature of the outer face would be much closer to the ambient air temperature.

Q13 (a) Peak emission $\lambda \sim 0.5\ \mu m$.

Using Wien's law: $= T = \dfrac{W}{\lambda_{peak}} = \dfrac{2.90 \times 10^{-3}\ m\ K}{0.5 \times 10^{-6}\ m} = 5800\ K = 6000\ K\ (1\ sf)$

(b) $\lambda_{peak} = \dfrac{W}{T} = \dfrac{2.90 \times 10^{-3}\ m\ K}{290\ K} = 1.0 \times 10^{-5}\ m = 10\ \mu m$

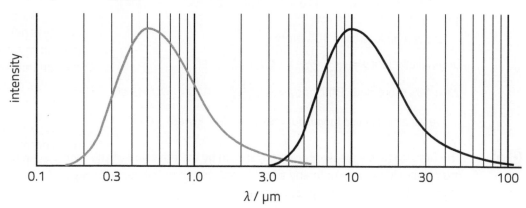

(c) Most of the solar radiation is between 0.4 and 0.9 µm, so reaches the surface of the Earth warming it up. The Earth's emitted radiation between 3 and 10 µm is partly absorbed by the atmosphere and re-emitted, 50% of which is downwards and warms the Earth further. The greater the concentration of the 'greenhouse gases', the greater the fraction of the Earth's radiation is absorbed and the higher the equilibrium temperature established – this is global warming.

Q14 (a) Height of tank $= \dfrac{volume}{csa} = \dfrac{400 \times 10^{-3}\ m}{\pi \times (0.30\ m)^2} = 1.415\ m$

Area = area of top + area of sides

$= \pi \times (0.30\ m)^2 + 2\pi \times 0.30\ m \times 1.415\ m$

$= 2.95\ m^2$

(b) The thermal conductivity of the steel is 1800× that of the PU and the thickness of the steel is much less (the surface areas of the steel and the PU are virtually the same).

(c) (i) $\dfrac{\Delta Q}{\Delta t} = 2.95\ m^2 \times 0.025\ W\ m^{-1}\ K^{-1} \times \dfrac{45^\circ C}{0.025\ m} = 133\ W$

(ii) $133\ W = 2.95\ m^2 \times 45\ W\ m^{-1}\ K^{-1} \times \dfrac{\Delta\theta}{0.003\ m}$

$\rightarrow \Delta\theta = 3 \times 10^{-3}\ ^\circ C\ \therefore$ justified.

[Alternatively: $45\ ^\circ C \times \dfrac{0.025}{45} \times \dfrac{0.3}{2.5} = 0.003\ ^\circ C,\ \therefore$ justified.]

(d) The heat transfer from the thermal store, albeit small, raises the temperature of the cupboard, thus reducing the temperature gradient and hence the power loss.

Component 3 Practice paper

1. (a) Longitudinal waves have oscillations in the same direction as the energy transfer whereas transverse waves have oscillations perpendicular to the energy transfer.

   (b) There is only one direction of oscillation for a longitudinal wave relative to energy transfer.

   (c) Direct light from the lamp is unpolarised (because there is no variation in intensity). The reflected light is partially polarised because there is a sinusoidal variation in the intensity, but the intensity never drops to zero. The unpolarised component of the reflected light is 30 arbitrary units. This is because a polaroid will always block 50% of the intensity of unpolarised light.

2. (a) 3rd row blanks: (7 **and** 1.4) or (7.3 **and** 1.46) or (7.3 **and** 1.5)
   6th row blanks: (36 **and** 3.6) or (35.7 **and** 3.57)

   (b) $v = \lambda f = 1.46$ cm $\times$ 3.57 Hz = 5.21 m s^{-1} [or 1.4 cm $\times$ 3.6 Hz = 5.0 m s^{-1}]

   (c) The readings in the table seem very imprecise – length is measured to the nearest cm rather than mm and the number of waves in 10 s has a large scatter in the values. A photograph of the waves (with a ruler alongside) will enable the length of five waves to be measured to the nearest mm – improving the precision by a factor of 10. A video of the moving waves could be analysed in slo-mo leading to an uncertainty of a small fraction of a wave in 10 s. Thus her claim seems realistic.

3. **Stationary waves**: Example – the oscillations of a string on a guitar. Energy is not propagated in the stationary wave. A series of nodes (zero amplitude) and anti-nodes (maximum amplitude) will be set up between the two ends of the string, with the amplitude varying between these two extremes. Between adjacent nodes, all the oscillations are in phase, with a jump of half a cycle when crossing a node.

   Progressive waves: Example – the sound wave of a singer. Energy is propagated from the source at the speed of sound. In the progressive wave, the amplitude will tend to decrease according to the inverse square law. The phase lag of the wave increases linearly as you move further from the source with one wavelength corresponding to a phase lag of 360° (or 2π radian).

4. (a) The waves travel different distances before arriving at some point. If the path difference (pd) is zero or a whole number of wavelengths then the waves will arrive in phase and constructive interference will occur. If the pd is an odd number of half-wavelengths then the waves will arrive in anti-phase and destructive interference will occur.

   (b) (i) Fringe separation $= \dfrac{9.2 \times 10^{-2}}{25} = 3.68 \times 10^{-3}$ m

   $\lambda = \dfrac{a\Delta y}{D} = \dfrac{0.25 \times 10^{-3} \text{ m} \times 3.68 \times 10^{-3}\text{m}}{1.800 \text{ m}} = 511$ nm

   (ii) This is due to diffraction at the grating slits. The diffraction pattern has a regular low intensity of diffraction at these points (this also suggests that the width of the slit is 4 times less than the slit separation).

(iii)

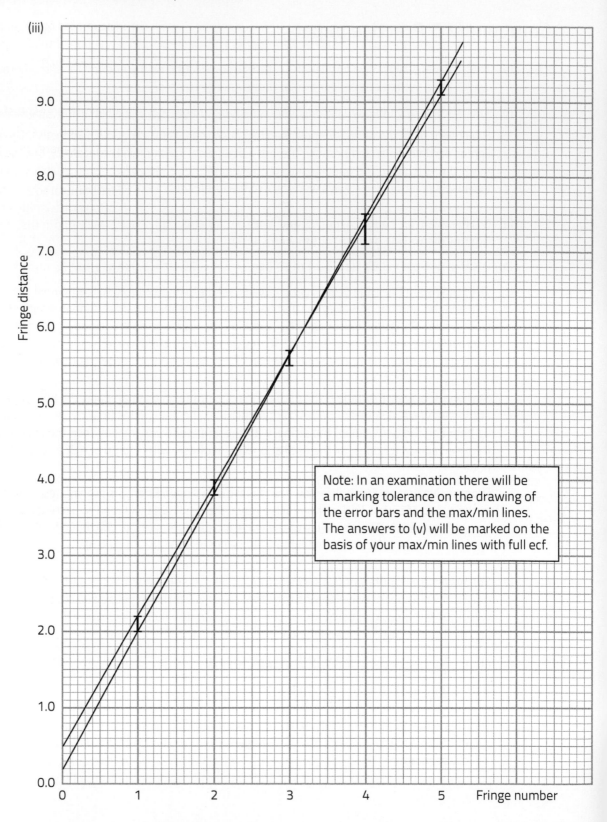

Note: In an examination there will be a marking tolerance on the drawing of the error bars and the max/min lines. The answers to (v) will be marked on the basis of your max/min lines with full ecf.

(iv) The max/min lines miss the origin. Probably due to locating the centre of the central fringe incorrectly (it could also be that a metre rule has been worn near the 0.0 cm mark leading to values that are systematically too large).

(v) Max gradient = $\dfrac{9.3 - 0.2}{25}$ = 0.364 [cm] ; Min gradient = $\dfrac{9.1 - 0.5}{25}$ = 0.344 [cm]

∴ Gradient = 0.354 ± 0.010 [cm] = (3.54 ± 0.10)× 10^{-2} [m]

3.54 ± 0.10 is a % uncertainty of 2.8%

$\lambda = \dfrac{a\Delta y}{D}$ and $\Delta y = \dfrac{y}{n}$ = gradient, so

∴ λ = gradient × $\dfrac{a}{D}$ = 3.54 × 10^{-2} m × $\dfrac{0.25 \times 10^{-3}\,\text{m}}{1.800\ \text{m}}$ (± 2.8%)

∴ λ = (4.92 ± 0.14) × 10^{-7} m or (4.9 ± 0.1) × 10^{-7} m

(vi) The fringes from a diffraction grating are sharper, brighter and separated by larger angles (because the slits are far closer together). This means that the location of the fringes themselves is more precise. Also, the greater separation of the fringes means that the percentage uncertainty in the angle (or distance) is much smaller.

5. (a) In spontaneous emission, an excited electron drops 'spontaneously', emitting a photon with energy equal to the energy gap of the energy levels involved. In stimulated emission, an excited electron is stimulated to drop by a photon with energy equal to the energy gap of the energy levels involved. The two resulting photons have the same phase, polarisation and direction.

    (b) A photon passing by the laser system can be absorbed or it can cause stimulated emission (or it can do nothing at all). In order to amplify the light, we need more stimulated emission than absorption. For more stimulated emission than absorption, we must have more electrons in the higher energy level than the lower energy level.

    (c) The 31.8 eV level is clearly the metastable level (longest lifetime) and must be the upper level in the laser emission.
    The difference in energy between the 'lasing' levels is $31.8 - 29.6 = 2.2$ eV
    Converted to J, this is $2.2 \times 1.6 \times 10^{-19} = 3.52 \times 10^{-19}$ J
    $E = \dfrac{hc}{\lambda}$, so $\lambda = \dfrac{hc}{E} = \dfrac{6.63 \times 10^{-34} \text{ J s} \times 3.00 \times 10^{8} \text{ m s}^{-1}}{3.52 \times 10^{-19} \text{ J}} = 565$ nm

6. (a) (i) $^{18}_{9}\text{F} \rightarrow ^{18}_{8}\text{O} + ^{0}_{1}\text{e (or e}^{+}) + \nu_e$

    (ii) B: $18 = 18 + 0 + 0$
    Q: $9 = 8 + 1 + 0$
    L: $0 = 0 - 1 + 1$

    (iii) Weak force

    (iv) The electron neutrino only feels the weak force. The lifetime is long. There is a change of quark flavour.

    (b) (i) $\lambda = \dfrac{\ln 2}{T_{1/2}} = \dfrac{\ln 2}{(109.8 \times 60) \text{ s}} = 1.052 \times 10^{-4} \text{ s}^{-1}$
    Then, using $A = \lambda N$
    $N = \dfrac{A}{\lambda} = \dfrac{360 \times 10^{6} \text{ Bq}}{1.054 \times 10^{-4} \text{ s}^{-1}} = 3.52 \times 10^{12}$
    Mass of a ^{18}F atom, $m(^{18}\text{F}) = 18 \text{ u} = 18 \times 1.66 \times 10^{-27}$ kg
    $\therefore$ Total mass $= 3.52 \times 10^{12} \times 18 \times 1.66 \times 10^{-27}$ kg
    $= 1.05 \times 10^{-13}$ kg

    (ii) Rearranging $A = A_0 e^{-\lambda t} \rightarrow e^{\lambda t} = \dfrac{A_0}{A}$
    Taking logs $\rightarrow t = \dfrac{1}{\lambda} \ln\left(\dfrac{A_0}{A}\right) = \dfrac{1}{1.052 \times 10^{-4} \text{ s}^{-1}} \ln\left(\dfrac{370}{20}\right)$
    $\therefore t = 28\ 000$ s (about 8 hours)

    (c) The initial momentum [of the electron and positron] is zero. Hence, by conservation of momentum, the momenta of the two photons must be equal and opposite. Photons with the same momentum have the same energy [$E = pc$].

7. (a) $^{11}_{5}\text{B}^{*} \rightarrow ^{7}_{3}\text{Li} + ^{4}_{2}\text{He}$

    (b) (i) Mass of RHS − LHS = $(10.01354 + 1.00867) - (7.01600 + 4.00260) = 0.00361$ u
    $\therefore$ Energy released $= 0.00361 \times 931 = 3.36$ MeV
    OR $0.00361 \times 1.66 \times 10^{-27} \times (3.00 \times 10^{8})^{2} = 5.39 \times 10^{-13}$ J

    (ii) The KE is only ($\sim 1/40$ eV) as stated in the question whereas the energy released is millions of times greater.

(c) Momentum is also conserved in the reaction. [The initial momentum of the boron–11 will be very small (or even negligible).] Hence, the momenta of the lithium and alpha particle will be equal and opposite after the decay.

$E_k = \frac{1}{2}mv^2 = \frac{1}{2}pv$ and since both nuclei have the same momentum, the particle with the greater speed has the greater KE.

(d) The stable boron–11 has a mass of 11.009 31 u which is less than the mass of the products [7.016 00 + 4.002 60 = 11.018 60] so would require an input of mass–energy for the reaction to occur.

(e) (i) Beta decay is not as ionising as alpha radiation; the ionisation will occur over a greater distance so that there is not enough concentration of ions to cause sufficient damage. [Plus the neutrino carries a lot of the energy release.]

(ii) [Issues question – a selection of the following points would suffice....]

Usually, isotopes are separated by using things like centrifuges which separate lighter and heavier isotopes. This is an expensive process involving high temperatures and multiple stages to gradually improve the purity. The boron would have to be melted (or maybe even evaporated). This would be a very expensive process to improve what is a pretty reliable process in the first place. The costs would never justify the improvements.

8. (a) (i) Using emf = $B\ell v$ = 0.15 T × 0.65 m × 8.0 m s^{-1} = 0.78 V

(ii) Choose your answer from:

Using FRHR, the current is upward, meaning that AB becomes positive.

OR, using FLHR on the electrons in the conductor, the electrons experience a downward force (meaning AB is positive again).

OR, the change is increasing the flux in the loop (YXAC), so any current in the loop will oppose this increase in flux. Using the right-hand grip rule (for the magnetic field of a loop), the current must flow anticlockwise (meaning the top becomes positive again).

(iii) An incomplete circuit means that no current flows and no resistive magnetic force can be set up.

(b) (i) $P = \dfrac{V^2}{R} = \dfrac{(0.78 \text{ V})^2}{0.50 \text{ }\Omega} = 1.22$ W

(ii) Power input = Fv

$\therefore F = \dfrac{P}{v} = \dfrac{1.22 \text{ W}}{8.0 \text{ ms}^{-1}} = 0.15$ N

(c) The component of the weight down the slope must provide the 0.15 N.

$\therefore mg \sin\theta = 0.15$ N

$\therefore \theta = \sin^{-1}\left(\dfrac{0.15 \text{ N}}{0.20 \text{ kg} \times 9.81 \text{ N kg}^{-1}}\right) = 4.4°$

Option A: Alternating currents

9. (a) $V_{max} = BAN\omega$

    $= 0.2\ \text{T} \times (5.0 \times 3.0) \times 10^{-4}\ \text{m}^2 \times 50 \times 2\pi \times 2.5\ \text{s}^{-1}$

    $= 0.24\ \text{V}$

    $T = 0.4\ \text{s}$

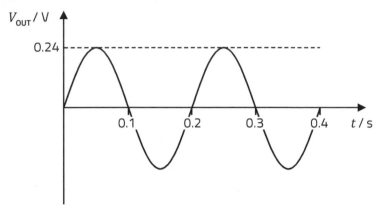

    (b) (i) The impedance, $Z = \sqrt{R^2 + (X_L - X_C)^2}$. At $3f_1$, the reactance of the inductor is 3×, i.e. 60 W and that of the capacitor is 1/3, i.e. 20 Ω. This gives the same impedance and hence the same current.

        (ii) (I) Current at resonance $= \dfrac{20\ \text{V}}{30\ \Omega}$

            $= 0.67\ \text{A}$

            If resonance frequency $= nf$.

            $\therefore \dfrac{60}{n} = 20n$

            $\therefore n^2 = 3 \rightarrow n = \sqrt{3} = 1.73$

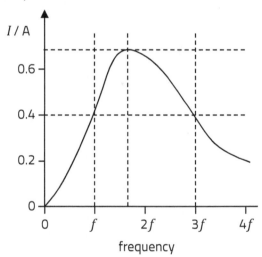

            (II) The Q factor can be calculated by $Q = \dfrac{\omega_0 L}{R}$. The values of ω_0 and L are unchanged, so if R is changed from 30 Ω to 50 Ω, the value of Q will be reduced (0.6×).

    (c) (i) Peak voltage $= \frac{1}{2} \times 6.0\ \text{div} \times 5\ \text{V div}^{-1} = 15.0\ \text{V}$

        Period $= 8.4\ \text{div} \times 50\ \mu\text{s div}^{-1} = 420\ \mu\text{s}$

        $\therefore$ Frequency $= 2381\ \text{Hz} = 2400\ \text{Hz}$ (2 sf)

        (ii) Peak $V_R = \frac{1}{2} \times 7.6\ \text{div} \times 2\ \text{V div}^{-1} = 7.6\ \text{V}$

            $\therefore$ Peak $I = \dfrac{7.6\ \text{V}}{150\ \Omega} = 0.051\ \text{A} \sim 50\ \text{mA}$

        (iii) $Z = \dfrac{V}{I} = \dfrac{15.0\ \text{V}}{0.051\ \text{A}} = 296\ \Omega = \sqrt{R^2 + X_L^2}$

            $\therefore X_L = \sqrt{296^2 - 150^2} = 255\ \text{W}$

            $\therefore L = \dfrac{X_L}{2\pi f} = \dfrac{255\ \Omega}{2\pi \times 2400\ \text{Hz}} = 0.017\ \text{H}$ (2 sf)

        (iv) From the CRO: phase difference $= \dfrac{1.4\ \text{div}}{8.4\ \text{div}} \times 2\pi$ (or 360°) $= 1.05\ \text{rad}$ (or 60°)

            From the R and X_L values:

            phase difference $= \tan^{-1}\left(\dfrac{X_L}{R}\right) = \tan^{-1}\left(\dfrac{255\ \Omega}{150\ \Omega}\right) = 1.04\ \text{rad}$ (or 59.5°)

            $\therefore$ These values are in agreement within CRO reading and rounding errors.

Option B: Medical physics

10. (a) (i) X-rays are emitted due to the deceleration of electrons as they strike the metal target.

(ii) From graph minimum wavelength = 25 pm

$$eV = \frac{hc}{\lambda} \rightarrow V = \frac{hc}{e\lambda} = \frac{6.63 \times 10^{-34}\,\text{J s} \times 3.00 \times 10^8\,\text{m s}^{-1}}{1.60 \times 10^{-19}\,\text{C} \times 25 \times 10^{-12}\,\text{m}} = 49.7\,\text{kV}$$

$P = IV = 0.055\,\text{A} \times 49\,700\,\text{V} = 2700\,\text{W}$

Efficiency = $\frac{37}{2700}$ = 0.014 = 1.4%

(iii)

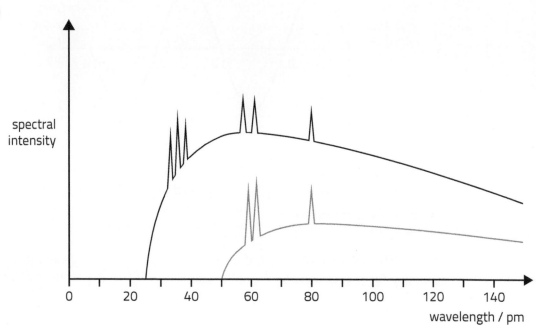

(b) $\frac{\Delta f}{f} = \frac{2v}{c}\cos\theta$. If ultrasound parallel to blood flow then q = 0 and cos q = 1

$$\therefore v = \frac{c\Delta f}{2f} = \frac{1570\,\text{m s}^{-1} \times 6.6 \times 10^3\,\text{Hz}}{2 \times 6.4 \times 10^6\,Hz} = 0.81\,\text{m s}^{-1}$$

(c) (i) The absorbed dose is the energy of radiation absorbed per unit mass for that organ. The equivalent dose for that organ is the absorbed dose multiplied by the radiation weighting factor.

(ii) Equivalent dose = 0.140 Gy × 20 = 2.80 Sv
Effective dose = 2.80 Sv × 0.12 = 0.34 Sv (2 sf)

(d) Ultrasound is non-ionising and can produce good images of soft tissue (although they are not the highest resolution). There will be a lot of reflections due to the bone (vertebrae) surrounding the spinal cord and the images are unlikely to be clear.

X-ray is high-resolution but has low soft-tissue contrast and involves low-dosage of ionising radiation. Additionally, the bones around the spinal cord will absorb the X-rays. The image is unlikely to be useful.

A CT scan has improved soft-tissue contrast but involves a dosage of ionising radiation. The absorption of X-rays by the vertebrae is likely to make the image one of poor quality.

MRI is ideal because of the good soft-tissue contrast and lack of ionising radiation. There is no problem with absorbing/reflecting bones affecting the image quality. Although MRI is usually double the cost of a CT scan it is best in this instance.

A PET scan is useless unless you are looking for cancer in the spinal cord. It also involves ionising radiation.

Option C: The physics of sports

11.   (a)   The double-decker bus has a much higher centre of gravity, **C**, compared to its width than an F1 car.

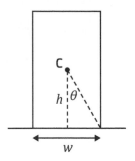

The rectangular object will topple if is tilted more than angle, q, where
$$\theta = \tan^{-1}\left(\frac{w}{2h}\right).$$
Hence θ is much greater for an F1 car than a double-decker bus.

   (b)   (i)   The rough side of the ball seems to hold the air stream longer than the polished side and the seam assists in separating the air stream from the ball on the right-hand side. These give the air a momentum to the right, meaning that the ball exerts a force on the air to the right (N2) and hence the air exerts an equal force to the left on the ball, which deviates to the left.

   (ii)   Consider the horizontal motion:

Time to travel 75 m, $t = \dfrac{75\,\text{m}}{(27.0\,\text{m s}^{-1})\cos 25°} = 2.806\,\text{s}$

Height as it reaches boundary,
$h = (27.0\,\text{m s}^{-1}) \times (2.806\,\text{s})\sin 25° - \frac{1}{2} \times (9.81\,\text{m s}^{-2}) \times (2.806\,\text{s})^2$
$= 2.10\,\text{m}$

So the batter is likely to be dismissed.

   (c)   (i) $I = mr^2 = 7.26\,\text{kg} \times (2.00\,\text{m})^2 = 29.0\,\text{kg m}^2$

   (ii)   Most of the mass of the hammer thrower is within 20 cm of the axis, so less than one tenth of the distance of the 7.26 kg ball. The contribution to the total moment of inertia depends on the square of the distance, so even with a 100 kg hammer thrower Σmr^2 would be much less than 29 kg m².

   (iii)   Using $\omega_2^2 = \omega_1^2 + 2\alpha\theta$ [the angular equivalent of $v^2 = u^2 + 2as$]

$\alpha = \dfrac{\omega_2^2 - \omega_1^2}{\theta}$. $\omega_2 = \dfrac{28\,\text{m s}^{-1}}{2.00\,\text{m}} = 14\,\text{rad s}^{-1}$. $\omega_1 = 0$

$\therefore \alpha = \dfrac{(14.0\,\text{rad s}^{-1})^2}{3.5 \times 2\pi\,\text{rad}} = 8.91\,\text{rad s}^{-2}$

$\therefore \langle\tau\rangle = I\alpha = (29.0\,\text{kg m}^2)(8.91\,\text{rad s}^{-2}) = 260\,\text{N m}$

Option D: Energy and the environment

12. (a) (i) The total power received by the Earth is $1360 \times \pi R^2$ W, where R is the radius of the Earth in m. The surface area of the Earth is $4\pi R^2$, so the power received by each m^2 of the Earth's surface is $1360\pi R^2 \div (4\pi R^2)$ W = 340 W.

    (ii) (I) Power emitted at 16 °C per m^2 = σT^4 = 5.67×10^{-3} W $m^{-2}K^{-4} \times (289$ K$)^4$ = 396 W.
    ∴ Additional power required = (396 − 340) W = 56 W.
    This is provided by radiation re-emitted from the atmosphere, mainly by the multi-atomic gases, water vapour, carbon dioxide and methane.

    (II) The thermal infra-red radiation emitted by the Earth's surface is partially absorbed by greenhouse gases and some is re-emitted towards the Earth's surface. With greater concentrations, more is absorbed and re-emitted, which increases the temperature of the surface of the Earth.

    (b) Breeding is the process of producing the fissile nuclide ^{239}Pu from the non-fissile ^{238}U. A ^{238}U nucleus (in a reactor fuel rod) absorbs a thermal neutron with the resulting {nuclide / ^{239}U} undergoing β^- decays to produce {a fissile nuclide / the fissile ^{239}Pu}. This increases the fuel in the reactor and enables more energy to be extracted from the fuel element than would otherwise be the case.

    (c) (i) Mass flow of water = 2.0×10^{-3} m^3 $s^{-1} \times 1000$ kg m^{-3}
    = 2.0 kg s^{-1}

    ∴ Heat flow per second = $mc\Delta\theta$ = 2.0 kg $s^{-1} \times 4200$ J kg°$C^{-1} \times (60 - 20)$°C
    = 336 000 W

    (ii) Mean temperature difference across the pipe wall = (70 − 40)°C = 30 °C
    If length of pipe = l, the surface area, A, of the pipe = $2\pi r l$

    Rate of heat transfer, $\dfrac{\Delta Q}{\Delta t} = AK\dfrac{\Delta\theta}{\Delta x}$

    ∴ 3.36×105 W = $2\pi l \times 1.0 \times 10^{-2}$ m $\times 385$ W m^{-1}°C $\times \dfrac{30\,°C}{1.8 \times 10^{-3}m}$

    = $4.03 \times 105\,l$ W m^{-1}

    ∴ l = 83 cm (2 sf)